An Introduction to Metascience

An Introduction to Metascience delves into core metascientific concepts, offering a critical examination of current knowledge creation processes and scrutinising researchers and their methodologies across disciplines.

This book stands alone as a comprehensive guide to metascience, offering readers a singular resource for understanding and implementing metascientific principles into their research practices. Readers will find this book invaluable for perfecting their research skills and enhancing the quality of their academic work. It exposes the reader to the intricacies of research processes, prompting a reevaluation of preconceived notions and fostering a deeper understanding of the flaws and solutions inherent in knowledge creation. Furthermore, it offers thought-provoking insights into implementing strategies to enhance research productivity, and it elucidates both the benefits and pitfalls of incorporating artificial intelligence in research production.

Designed for scientists and researchers seeking to gain insight into the scientific process, *An Introduction to Metascience* caters to those interested in understanding how research evolves over time. It appeals to individuals eager to explore methods, practices, and philosophies of science to refine their approach to knowledge creation.

Dr Gabriel Bennett, the pen name for Dr Matthew Bennett, holds a PhD in Disability Studies from Flinders University, Australia. He has lectured in Disability Studies at Griffith University, Queensland. He has also advised the Australian Government's Autism CRC and has published articles for the *Journal of Autism and Developmental Disorders*. He is actively involved in supporting autistics to achieve their potential in society by disseminating his knowledge about the autism spectrum via lectures, conference presentations, and publications.

Dr Emma Goodall is an adjunct research fellow at the University of Southern Queensland, Australia, holds a PhD in Education, and is focused on teaching students on the autism spectrum. She is an executive member of the Australian Society for Autism Research, independent researcher through Healthy Possibilities, and a published author and keynote speaker in the areas of autism, sexuality and relationships, education, and interoception.

An Introduction to Metascience

The Discipline of Evaluating the Creation
and Dissemination of Research

Gabriel Bennett and Emma Goodall

LONDON AND NEW YORK

Cover image: Getty Images © Artemisia1508

First published 2025
by Routledge
4 Park Square, Milton Park, Abingdon, Oxon OX14 4RN

and by Routledge
605 Third Avenue, New York, NY 10158

Routledge is an imprint of the Taylor & Francis Group, an informa business

British Library Cataloguing-in-Publication Data
A catalogue record for this book is available from the British Library

Library of Congress Cataloging-in-Publication Data
Names: Bennett, Gabriel, author. | Goodall, Emma, 1971– author.
Title: An introduction to metascience : the discipline of evaluating the creation and dissemination of research / Dr. Gabriel Bennett, PhD, Dr. Emma Goodall, PhD.
Description: Abingdon, Oxon ; New York, NY : Routledge, 2025. |
Includes bibliographical references and index.
Identifiers: LCCN 2024021769 | ISBN 9781032769066 (pbk) |
ISBN 9781032836386 (hbk) | ISBN 9781003510376 (ebk)
Subjects: LCSH: Science—Methodology. | Science—Philosophy.
Classification: LCC Q175 .B456 2025 | DDC 507.2—dc23/eng/20240617
LC record available at https://lccn.loc.gov/2024021769

ISBN: 9781032836386 (hbk)
ISBN: 9781032769066 (pbk)
ISBN: 9781003510376 (ebk)

DOI: 10.4324/9781003510376

Typeset in Times New Roman
by codeMantra

We dedicate this book to future researchers worldwide who will spend their careers advancing humanity's understanding of different academic disciplines.

Contents

1 An introduction to metascience

Abstract

This chapter begins with a broad description of metascience, including its various aspects and why a proficient understanding and application of metascience strengthens the quality of academic research. This is followed by sections that describe the intended audience for this book, the pedagogical features in this book that will help different readers comprehend the diverse aspects of metascience, and an outline of the contents of the upcoming chapters.

1.1 What is metascience?

During 1996, Alan Sokal created the now-famous 'Sokal hoax' in which he published in the journal *Social Text* the fake paper "Transgressing the Boundaries: Towards a Transformative Hermeneutics of Quantum Gravity". He did this hoax to evaluate if a nonsensical article that appealed to the ideological prejudices of the editor would be published. Sokal's article was not scrutinised by quantum physicists because at the time *Social Text* published studies that did not undergo peer review. Three weeks later *Lingua Franca* published Sokal's claim that the article he had submitted to *Social Text* was bogus. It is easy to dismiss Sokal's actions as nothing more than a stunt. However, they have raised questions and concerns about the potential influence that academic papers can have on editorial decisions (Secor & Walsh, 2004).

It should have been a wake-up call for academic publishers when Sokal admitted that he had been able to successfully publish a fake article. However, instead of taking the opportunity to reform and safeguard the integrity of their academic journals, publishers either ignored or condemned Sokal for his hoax. Their failure to acknowledge and mitigate the weaknesses that Sokal exposed has given others the opportunity to conduct similar publishing hoaxes. Another successful hoax occurred from 2017 to 2018 when Helen Pluckrose, James Lindsay, and Peter Boghossian submitted 20 fake articles to a range of academic journals that specialised in publishing intersectional feminism, critical race theory, queer theory, postcolonial theory, and fat studies. Their publishing hoax came to be widely known as the 'grievance studies affair' or alternatively, but less commonly, 'Sokal squared' (Pluckrose et al., 2021).

The grievance studies affair had two stages. The first stage involved Lindsay, Boghossian, and Pluckrose submitting six fraudulent manuscripts to academic journals. These manuscripts were completely incoherent and lacked any rigour or logical connections to the established body of academic research. Put simply, 'these papers were, in short, nonsense sprinkled, mostly irrelevantly, with jargon' (Pluckrose et al., 2021, p. 1919). However, this approach did not result in any publications. Instead, just one fake manuscript was sent out by the academic journal *Men*

DOI: 10.4324/9781003510376-1

and Masculinity for peer review and then rejected. Based on this outcome, Lindsay, Boghossian, and Pluckrose speculated that:

> Had these papers been published, it would have demonstrated that there was a serious problem with peer review for the targeted journals and that editors and reviewers were accepting papers without properly reading and understanding them.
>
> (Pluckrose et al., 2021, p. 1919)

Pluckrose, Lindsay, and Boghossian regard the first stage to be the 'hoax phase', which closely resembled Sokal's actions (i.e., both Sokal and the first stage of the grievance studies affair involved academic journals receiving bogus papers). After receiving an overwhelming desk rejection rate, Pluckrose, Lindsay, and Boghossian realised that academic journals would not accept manuscripts that were just pure nonsense. Thus, they began the second stage of the grievance studies affair, which involved combining established scholarship with a lack of empirical rigour, rejection of any logical reasoning, and a substantial application of discriminatory identity theory and authoritarian intolerance of alternative perspectives. Thus, instead of trying to entice academic journals into publishing manuscripts that they viewed as unappealing, which occurred in stage one, Pluckrose, Lindsay, and Boghossian used the opposite approach. For stage two, they attempted to get their manuscripts to the peer review stage by writing them in a way that appealed to the editor. If they were successful, then they allowed the peer reviewers themselves to guide their manuscript's development to publication. Technically, the second stage does not fit the strict definition of a 'hoax' because the academic journals published the manuscripts based on their criteria for peer review. This shift in approach, from submitting to journals nonsense articles that were not appealing to submitting articles that appealed to a journal's ideological biases, yielded significant results. Of the 14 manuscripts submitted for stage two all but one manuscript was sent out for peer review (Pluckrose et al., 2021).

After the grievance studies affair was revealed, it received mixed responses from scholars. Some have praised this affair because it exposed what they believed were widespread methodological flaws within some sectors of the humanities and social sciences. For example, Yascha Mounk claimed that their actions revealed that 'some of the leading journals in areas like gender studies have failed to distinguish between real scholarship and intellectually vacuous as well as morally troubling bullshit' (Singal, 2019). In contrast, others have criticised this affair. For example, Geoff Cole (2021) has argued that knowledge could have been inevitably corrupted since some of the hoax studies that Pluckrose, Lindsay, and Boghossian attempted to get published contained false data (i.e., data fabrication). In a rebuttal to this accusation, Pluckrose, Lindsay, and Boghossian argued that:

> We liken our approach to that of a 'white hat' investigation in which individuals attempt to get fake passports through border control, or hack a security system, in order to test whether systems that we should be able to rely on are actually up to the job. Quality assurance and penetration tests of this kind are standard in many professional settings. If the sanctity of the scientific record is the primary concern, then surely the benefits of revealing the acceptance of preposterous data and gross mishandling of it by academic journals we rely on for scholarship into issues of social outweighs a temporary fabrication of such data. It was always our intention to reveal our project so no permanent corruption of the scientific record could possibly ensue. In that regard—and as we absolutely endorse the

retraction of all of our papers that utilized fabricated data (with proper explanation as such)—we stand by our method.

(Pluckrose et al., 2021, pp. 1925–1926)

A typical, and somewhat predictable, reaction among some in academia is to condemn Pluckrose, Lindsay, Boghossian, and Sokal. Undoubtedly, some who have devoted their entire careers to working in academia believe they were 'bomb throwers' who were sabotaging the creation and purity of knowledge or 'jokers and time wasters' who were stealing a peer reviewer's precious time. Such ad hominem criticisms ignore the potentially serious flaws in the creation of research whilst simultaneously attacking those who prove such flaws exist. Continuing to 'play the man and not the ball' is an unsustainable means of preserving the creditability of academia, especially since high-profile cases of academic fraud continue to occur (Miles, 2022) and article retention rates have not decreased.

The hoax papers that Pluckrose, Lindsay, Boghossian, and Sokal have published should be regarded as the 'tip of the iceberg' of what is a flawed system for creating research. If we look under the ocean's surface, at the entire iceberg, we will see that there are many factors that undermine the creation of credible research. For example, a grant agency's priorities about what research they will fund, the culture of research workplaces and the dynamics between colleagues, and the conduct of ethics committees can all influence what, how, and the quality of research that is produced. Additionally, there are many problems in our systems when creating research. For example, authors are rarely obligated to give journals a copy of their study's dataset when they submit their manuscript for peer review. Consequently, peer reviewers are unable to repeat the same analysis using the same dataset to confirm if the reported results are truthful, a process known as 'reproducibility' (i.e., see Chapter 7). The lack of reproducible research has meant that potentially flawed research has passed peer review and is now deemed to be credible scholarship.

To safeguard the integrity of scholarship, those who create knowledge and the tools and processes that they use to conduct this activity need to be studied and improved. Metascience, otherwise known as meta-research, is the discipline of examining the processes used to create research and the incentives, motivations, and actions of researchers when they are producing knowledge. According to Khakshooy et al. (2020, p. 5) 'metascience involves the use of new and stringent scientific methodology to study science itself for raising the overall quality of scientific knowledge'. They also claim that 'the very goal of metascience is to ensure that scientific progress and information grow from accurate, systematically verified, statically incontrovertible, and unquestionably true facts' (Khakshooy et al., 2020, p. 5). As Ioannidis et al. (2015, p. 3) colloquially explained metascience 'involves taking a bird's eye view of science'. Thus, metascience is both the study of the conduct and motives of scholars and the processes and tools that they use to create and disseminate research. Although not a complete description, Ioannidis et al. (2015) have proposed that metascience is composed of five distinct aspects, which are methods, reporting, evaluation, reproducibility, and incentives (see Table 1.1).

Along with Ioannidis et al., Roberts and PLOS Biology Staff Editors (2022) have also illustrated the main components of metascience and how they are all interconnected (see Figure 1.1). The components of their illustration are now explained:

1 *Composition and behaviour of the research community:* The production of research is influenced by those undertaking this activity. *PLOS Biology* have published studies that have investigated how the research topic as well as the researcher's gender and career stage might impact collaboration activities (Zeng et al., 2016) and how long it might take for men and women to be equally represented in different academic disciplines (Holman et al., 2018).

Table 1.1 Five major aspects of metascience

Metascience subdiscipline	Specific interests (non-exhaustive list)
Methods: "performing research" – study design, methods, statistics, research synthesis, collaboration, and ethics	Biases and questionable practices in conducting research, methods to reduce such biases, meta-analysis, research synthesis, integration of evidence, crossdesign synthesis, collaborative team science and consortia, research integrity and ethics
Reporting: "communicating research" – reporting standards, study registration, disclosing conflicts of interest, information to patients, public, and policy makers	Biases and questionable practices in reporting, explaining, disseminating and popularising research, conflicts of interest disclosure and management, study registration and other bias prevention measures, and methods to monitor and reduce such issues
Reproducibility: "verifying research" – sharing data and methods, repeatability, replicability, reproducibility, and self-correction	Obstacles to sharing data and methods, replication studies, replicability and reproducibility of published research, methods to improve them, effectiveness of correction and self-correction of the literature, and methods to improve them
Evaluation: "evaluating research" – prepublication peer review, post-publication peer review, research funding criteria, and other means of evaluating scientific quality	Effectiveness, costs, and benefits of old and new approaches to peer review and other science assessment methods, and methods to improve them
Incentives: "rewarding research"– promotion criteria, rewards, and penalties in research evaluation for individuals, teams, and institutions	Accuracy, effectiveness, costs, and benefits of old and new approaches to ranking and evaluating the performance, quality, value of research, individuals, teams, and institutions

Source: Ioannidis et al. (2015, p. 3).

2   *Choice of research topics:* The types of research conducted are influenced by the selection of participants, animals, tissues, cells, or genes for study. For example, one study published in *PLOS Biology* investigated what factors influenced the volume of resources devoted to exploring each of the 19,000 human genes (Stoeger et al., 2018).

3   *Design and methodology of research studies:* Once a researcher has chosen a topic, the study and the subsequent analysis can be carried out in a variety of different ways. For example, *PLOS Biology* has published a study about the advantages of sample heterogeneity in pre-clinical work (Voelkl et al., 2018).

4   *Description and interpretation of results:* Typically, a study's results are placed into a standardised manuscript structure (i.e., abstract, introduction, methods, results, discussion, conclusion, and references). Metascientists have investigated the effect of sensationalising findings (Chiu et al., 2017) and the methods used to prepare pictures for publication (Jambor et al., 2021).

5   *Publication and dissemination of results:* A study's posting on a preprint server, press coverage, and social media exposure can influence the direction of research and the public's understanding of knowledge. For example, *PLOS Biology* has published studies about the usage of scientific preprints by right-wing social media groups (Carlson & Harris, 2020) and the impact of article titles in press coverage (Triunfol & Gouveia, 2021).

6   *Assessment of researchers and funding decisions:* Among other considerations, agencies that allocate grants for research typically use a researcher's publications to determine if they should receive a research grant. Their funding decisions can influence the types of studies that are conducted, with researchers choosing to undertake studies that are likely to be

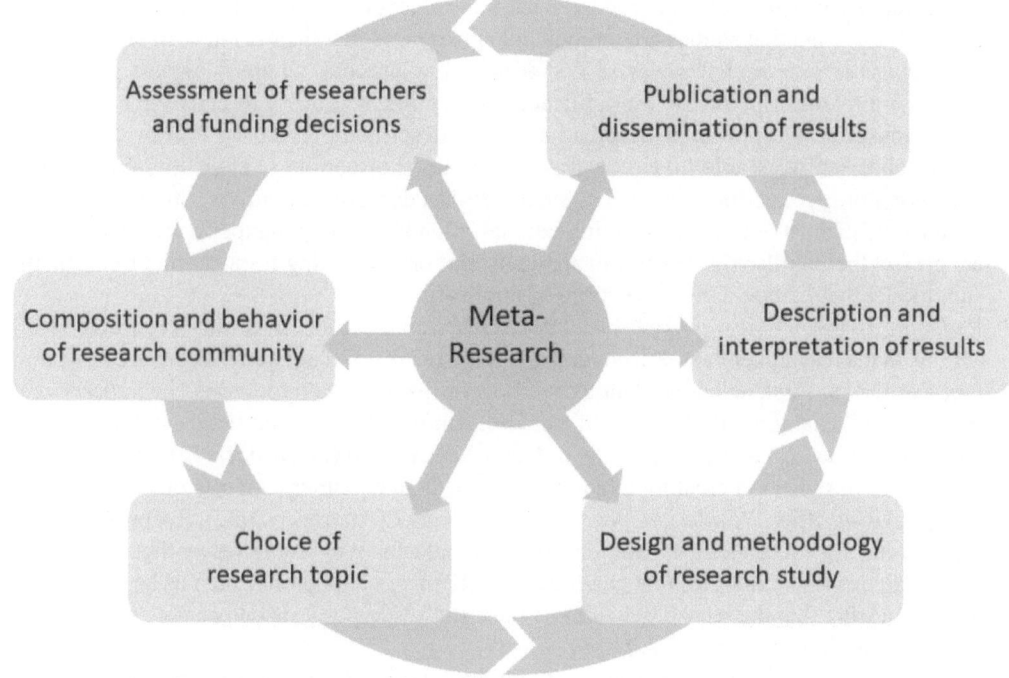

Figure 1.1 Metascience interrogates all stages of the research lifecycle
Source: Roberts and PLOS Biology Staff Editors (2022, p. 2).

published instead of studies that would be more difficult to publish but would yield greater benefits to society. *PLOS Biology*, for example, has published studies to quantify a publication's impact (Hutchins et al., 2016) and innovative approaches for allocating research funds (Gross & Bergstrom, 2019).

Unfortunately, tracing the historical origins of metascience is exceptionally difficult since its concepts have not been combined to form a unique discipline. It can be argued, however, that metascience is a relatively new movement; starting with John Ioannidis's studies and opinion pieces about it since 2005 (Ioannidis, 2005). Despite metascientific concepts being dispersed throughout different disciplines, the trend of amalgamating metascientific principles and concepts into a coherent field has only recently begun. An inaugural conference on metascience was held in 2019, followed up in 2021 by a second metascience conference (Metascience Symposium, 2019, 2021).

The main goal of any metascientist is to evaluate and improve the production of research. To do this, they are likely to perform most or all the following tasks:

1 *Ensuring that the results in a study can be confirmed*: For a study's results to be deemed as credible identical results need to be created after a reanalysis of the same dataset using the same methodological tools and procedures has occurred. The act of repeating this process is called 'reproducibility'. If it is not possible to perform this task, either because the dataset is not published and/or the methodological tools or procedures are inadequately described, then the study

is deemed 'irreproducible'. Consequently, the truthfulness of the study's results will always be in doubt. Ensuring that studies are reproducible guarantees that their results can be confirmed.

2   *Improving the peer review process*: The peer review process, whereby experts review and decide if a manuscript should be published, is a common process to certify the credibility of a manuscript. Despite its widespread usage, as described previously, Pluckrose, Lindsay, and Boghossian were able to demonstrate that it has shortcomings. Thus, in the interests of improving the production of research, metascientists explore and purpose strategies that can enhance it, such as developing training courses intended to help inexperienced peer reviewers perfect their skills with examining a manuscript or advocating for a 'results free' submission model that hampers the occurrence of publication bias.

3   *Reducing questionable research practices (QRPs)*: QRPs are generally regarded as any actions that a researcher can perform that alters the results of their study. There are three main types of QRPs. First, selecting data that proves or disproves a hypothesis (i.e., *cherry picking*). Second, changing the contents of a dataset and/or selecting a particular statistical test to create a *p*-value that is above or below a statistical threshold (i.e., *p-hacking*). Third, after the data has been collected creating a hypothesis that the data either supports or disproves (i.e., *Hypothesising After Results are Known* – HARKing) (Andrade, 2021). To prevent QRPs from occurring metascientists examine the rationale for why some researchers use QRPs. They also develop strategies that can prevent QRPs from happening, such as academic journals requesting that the researcher submits their study's hypothesis before they conduct their study so that they cannot do HARKing.

4   *Improving the employment prospects of suitably qualified and experienced research candidates*: Sometimes, people from minority groups (e.g., women of colour) are unable to obtain academic opportunities despite having suitable skills, experience, and qualifications. Such employment barriers undermine the workplace's efficiency and hamper the production of high-quality research since the most capable candidate is not employed. Metascientists examine hiring practices and propose strategies to ensure that the most suitable candidate is awarded the academic position regardless of their immutable characteristics, such as their ethnicity, gender expression, age, religious affiliation, and sexual orientation.

5   *Enhancing the productivity of research workplaces*: Like other workplaces, bullying and discrimination happen in some research workplaces. Such behaviours can reduce productivity since staff either resign or take leave. Metascientists examine the professional conduct of researchers so that bullying and other behaviours cannot reduce the productivity of research workplaces.

6   *Maximising the financial investment spent on research*: Depending on the time and resources required, conducting research can be expensive. Metascientists examine ways that research activities can be conducted more efficiently to ensure that resources are efficiently allocated.

7   *Protecting participants by improving ethics application and review processes*: Sometimes, members of ethics committees do not have the expertise and knowledge to proficiently evaluate an ethics application. Consequently, they might make incorrect judgements, such as rejecting a study despite its benefits outweighing its risks to participants. In the interests of protecting the dignity and safety of participants, metascientists examine the shortcomings of the ethics application process and recommend ways that it can be improved.

8   *Improving academic recruitment processes*: The number of publications and the prestige of the academic journal where they were published are often used as a proxy to determine the most suitable candidate for an academic grant or university position. Other academic tasks, such as peer reviewing potential journal articles, are rarely given the same consideration.

Some metascientists have advocated for a more comprehensive approach when making such recruitment decisions. A consequence of this proposed approach is that academics with a broader academic skill set, compared to those who are just competent at publishing studies, will be appointed and the practice of scholarship would subsequently improve.

9 *Enhancing the study's design*: Metascientists examine a study's inherent drawbacks and benefits. Failure to identify a study's drawbacks can result in the creation of low-quality research that is both a waste of money and human resources. To avoid this outcome, metascientists evaluate a study's design with the intention of correcting any shortcomings. For example, as explained in the next chapter, a metascientist might propose strategies to mitigate the detrimental influences that biases can have on a study's results.

10 *Increasing the dissemination of research and its results*: Since new discoveries often stem from previous discoveries, it is important that a study's results are disseminated to as many scholars as possible. Failure to widely disseminate a study's findings can result in scholars being unaware of relevant developments in their discipline and subsequently conducting irrelevant or repetitive research. Metascientists design strategies that promote research dissemination so that scholars are more effectively informed about developments in their field.

11 *Refining the usage of artificial intelligence and machine learning*: Arguably, the introduction of machine learning and artificial intelligence is one of the most recent and profound technological developments that has assisted scholars in the production of research. For example, machine learning has reduced the time scholars use to identify suitable studies for literature reviews as well as assist with both quantitative and qualitative data analysis. In the interests of ensuring that machine learning programmes are flawless metascientists study how information technology can be improved to produce more accurate results.

Despite its benefits, there are two plausible reasons why there is a lack of willingness among researchers to learn about metascience and to apply its principles. First, metascientific concepts and principles are scattered among individual disciplines. For example, the metascientific concept of 'reproducibility' has been applied to the study of social sciences (Hardwicke et al., 2020), psychology (Hardwicke et al., 2021), and neurological research (Rauh et al., 2020). Furthermore, the academic journal *Research Integrity and Peer Review* only publishes manuscripts about the reliability of the peer review process, editorial decision making, and research and publication ethics (Harriman et al., 2016). By synthesising metascientific concepts, arguably, this book will help formalise the discipline of metascience. Second, typically scholars are focused on creating and then disseminating new knowledge via publications, conference presentations, and winning research grants. Consequently, there is no incentive for them to stop their production of knowledge so that they can evaluate and improve the quality of their research using metascientific concepts. Furthermore, metascience can also put some researchers into an uncomfortable position whereby they realise that their research might not be as rigorous and as accurate as they previously believed.

1.2   The intended audience of this book

Despite being an obscure discipline, the concepts and principles of metascience presented in this book can be of value to different audiences. First, academics can use this book to assist them when explaining to their student's fundamental aspects of research production, such as preserving datasets to either assist with data collected for other studies or ensure that a study's results can be validated (i.e., reproducibility, see Chapter 7). Second, this book can help university administrators improve the workplace for researchers. For example, administrators can use the ideas outlined in Chapter 5 to produce equitable and inclusive research workplaces. Third, PhD candidates and researchers can use the principles and concepts in this book to improve their research practices. Fourth, grant funding agencies can implement the suggestions outlined in Chapter 4 to improve the feedback that they provide to unsuccessful candidates.

1.3   Pedagogical features in this book

Students learn in different ways and to cater for these different learning styles this book contains a variety of pedagogical features to help emphasise key concepts, promote better information retention, and consolidate what the reader has learned. This book contains the following features:

- *Lists of chapter objectives:* A list of key chapter objectives will be presented at the start of the chapter. Its purpose is to give the reader a quick snapshot of the main concepts in the chapter and what they will learn.
- *Diagrams or illustrations:* Some readers learn concepts via diagrams and/or illustrations. When a concept can be explained visually an illustration or diagram will be provided.
- *Checklists:* This book contains a series of checklists that the reader can use to improve both their production and examination of studies. For example, the *Sex and Gender Equity in Research* (SAGER) guidelines, a checklist for gender-sensitive reporting, and a series of checklists by JBI for examining different academic pieces are presented.
- *Further reading:* At the end of each chapter a list of further readings relevant to the concepts discussed will be listed. The purpose of this list is to give the reader additional insights about the concepts explored.

1.4   Summary of the upcoming chapters

1.4.1   Chapter 2 – Mitigating biases during the production and dissemination of research

Biases can influence how a study is designed and conducted, and they can hamper the production of reliable results. When one biological sex is prioritised in research, for instance, gender bias happens during the study's design phase. Gender bias has affected studies about the COVID-19 pandemic and the autistic spectrum. Biases can also occur during the examination of research. For instance, publication bias occurs when studies with results that journal editors find appealing are chosen for evaluation and when peer reviewers advise publishing a study due to its components (e.g., aims, methodology, and results). Finally, biases can emerge during the research dissemination phase. For instance, an internet search engine's algorithmic architecture dictates the results that it generates. Those who read this chapter should be able to answer the following questions:

- What are some common biases that might happen during the design, data collection, and dissemination phases of the research process?
- What strategies can be implemented to reduce the occurrences of biases?

1.4.2   Chapter 3 – Journalology: the science of publishing

Once finished, researchers typically publish their study and its results in peer-reviewed publications. In Chapter 3 common problems and potential solutions with the peer review process are discussed. The negative impacts of predatory journals and preprinted manuscripts along with strategies to reduce their negative impacts are also explained in this chapter. The information in this chapter should give the reader the ability to respond to the following queries:

- What are some common problems and potential solutions to the peer review process?
- What techniques can a researcher use to identify an article published by a predatory publisher?
- What checks can a researcher perform to identify publication invites from predatory publishers?
- What methods can publishers use to guarantee that readers are made aware that they are reading a preprinted manuscript?

1.4.3   *Chapter 4 – The impact of funding agencies on the production of research*

The rules that funding organisations apply to their funding opportunities can have an impact on the conduct of research teams. For example, Diong et al. (2021) examined the mandates for reproducible research practices that eight funding opportunities in Australia placed on applicants. They reported that no funding agency either 'required' or 'encouraged' the applicant to publicly register their study protocols before they started data collection or that they had to make their datasets publicly available once the study was finished. This lack of conditions meant that other researchers were unable to confirm the results by repeating the same analysis of the same dataset. Although upcoming researchers may have more deserving research topics, current submission procedures can entrench and favour existing researchers. To ensure newer generations can access research funding, grant agencies should use a two-stage application process and reserve some grant funding for only applicants who have received their doctorate in the previous five years. This chapter can help the reader answer the following questions:

- What guidelines and processes should agencies implement or change to improve their decisions about what studies they will fund?
- How can a grant reviewer improve their assessment of the research proposal's financial merits?

1.4.4   *Chapter 5 – Improving the culture in research workplaces*

Some research staff have reported bullying, racism, and sexism in the workplace. When such discriminatory behaviour happens the quality of their academic research may be compromise. To maintain the integrity of research activities within Chapter 5 strategies that can inhibit the occurrence of bullying, racism, and sexism are outlined. Those who read this chapter should be able to answer the following questions:

- How does racism within academic workplaces prevent people from ethnic minorities from becoming academics?
- What modifications can be implemented to overcome a male-focused recruitment bias and ensure that recruitment processes for academic positions are meritocratic?
- How can research institutions improve the representation and participation of academics from underrepresented backgrounds?

1.4.5   *Chapter 6 – Understanding and addressing QRPs*

Some researchers have used questionable research practices (QRPs) to increase the likelihood of their study getting published, which subsequently benefits their career and grant winning prospects.

QRPs have the potential to undermine public faith and investment in research. To prevent this outcome Chapter 6 introduces the reader to three common QRPs and solutions that can stop them from occurring. Those who read this chapter will be able to answer the following questions:

- What is a QRP?
- What factors motivate researcher to use QRPs?
- What techniques can be used to prevent researchers from using QRPs?

1.4.6   *Chapter 7 – Addressing the reproducibility crisis*

To validate the findings of a study, identical results must be generated after using the same analytical processes on the same dataset. Despite the importance of authenticating a study's results it is standard practise to destroy datasets once a study has been published. This custom has created a 'reproducibility crisis'. Chapter 7 begins with a description of reproducibility and the extent of irreproducible research within different disciplines. The economic and social implications of irreproducible research are then outlined. To mitigate the impact of these consequences strategies that can increase reproducible research are then outlined, including removing obstacles to making datasets available and establishing journals that only publish reproducible studies. The purpose of this chapter is to improve the likelihood of reproducible research being produced, which will ensure that the truthfulness of results can be protected. Those who read this chapter should be able to answer the following questions:

- What are the social and financial consequences of creating irreproducible research?
- What strategies can be used to improve the reproducibility of research?

1.4.7   *Chapter 8 – Ethics and metascience*

Before doing any human research, a researcher must first acquire ethical approval from their institution's Human Research Ethics Committee (HREC). Although applying for ethics clearance is a straightforward process, HRECs are themselves not flawless. Some HREC members, for example, may be confused about the ethical implications in the ethics proposal because they are unfamiliar with the scientific discipline. This, as well as other typical weaknesses of HRECs, are discussed in Chapter 8. This chapter is intended to help newly minted HREC members understand their duties and responsibilities in HRECs. The After reading this chapter, the following questions could be answered by the reader:

- What impacts do the views and prejudices of members of HRECs have on their decisions to provide ethics approval?
- After a study is finished should HRECs conduct a 'retrospective ethics review', in which the researcher informs the HREC about ethical concerns that the HREC did not anticipate when granting initial ethics approval?
- How can HRECs be reformed so that they can competently examine ethics applications that contain multidisciplinary and/or highly specialised concepts?

References

Andrade, C. (2021). HARKing, Cherry-picking, P-hacking, fishing expeditions, and data dredging and mining as questionable research practices. *The Journal of Clinical Psychiatry, 82*(1), 20f13804. https://doi.org/10.4088/JCP.20f13804

Carlson, J., & Harris, K. (2020). Quantifying and contextualizing the impact of bioRxiv preprints through automated social media audience segmentation. *PLoS Biology, 18*(9), e3000860. https://doi.org/10.1371/journal.pbio.3000860

Chiu, K., Grundy, Q., & Bero, L. (2017). 'Spin' in published biomedical literature: A methodological systematic review. *PLoS Biology, 15*(9), e2002173. https://doi.org/10.1371/journal.pbio.2002173

Cole, G. G. (2021). Why the "Hoax" Paper of Baldwin (2018) should be reinstated. *Sociological Methods & Research, 50*(4), 1895–1915. https://doi.org/10.1177/0049124120914951

Diong, J., Kroeger, C. M., Reynolds, K. J., Barnett, A., & Bero, L. A. (2021). Strengthening the incentives for responsible research practices in Australian health and medical research funding. *Research Integrity and Peer Review, 6*(1), 11. https://doi.org/10.1186/s41073-021-00113-7

Gross, K., & Bergstrom, C. T. (2019). Contest models highlight inherent inefficiencies of scientific funding competitions. *PLoS Biology, 17*(1), e3000065. https://doi.org/10.1371/journal.pbio.3000065

Hardwicke, T. E., Bohn, M., MacDonald, K., Hembacher, E., Nuijten, M. B., Peloquin, B. N., deMayo, B. E., Long, B., Yoon, E. J., & Frank, M. C. (2021). Analytic reproducibility in articles receiving open data badges at the journal *Psychological Science*: an observational study. *Royal Society Open Science, 8*(1), 201494. https://doi.org/10.1098/rsos.201494

Hardwicke, T. E., Wallach, J. D., Kidwell, M. C., Bendixen, T., Crüwell, S., & Ioannidis, J. (2020). An empirical assessment of transparency and reproducibility-related research practices in the social sciences (2014–2017). *Royal Society Open Science, 7*(2), 190806. https://doi.org/10.1098/rsos.190806

Harriman, S. L., Kowalczuk, M. K., Simera, I., & Wager, E. (2016). A new forum for research on research integrity and peer review. *Research Integrity and Peer Review, 1*, 5. https://doi.org/10.1186/s41073-016-0010-y

Holman, L., Stuart-Fox, D., & Hauser, C. E. (2018). The gender gap in science: How long until women are equally represented? *PLoS Biology, 16*(4), e2004956. https://doi.org/10.1371/journal.pbio.2004956

Horbach, S., Breit, E., Halffman, W., & Mamelund, S. E. (2020). On the willingness to report and the consequences of reporting research misconduct: The role of power relations. *Science and Engineering Ethics, 26*(3), 1595–1623. https://doi.org/10.1007/s11948-020-00202-8

Hutchins, B. I., Yuan, X., Anderson, J. M., & Santangelo, G. M. (2016). Relative citation ratio (RCR): A new metric that uses citation rates to measure influence at the article level. *PLoS Biology, 14*(9), e1002541. https://doi.org/10.1371/journal.pbio.1002541

Ioannidis, J. P. (2005). Why most published research findings are false. *PLoS Medicine, 2*(8), e124. https://doi.org/10.1371/journal.pmed.0020124

Ioannidis, J. P., Fanelli, D., Dunne, D. D., & Goodman, S. N. (2015). Meta-research: Evaluation and improvement of research methods and practices. *PLoS Biology, 13*(10), e1002264. https://doi.org/10.1371/journal.pbio.1002264

Jambor, H., Antonietti, A., Alicea, B., Audisio, T. L., Auer, S., Bhardwaj, V., Burgess, S. J., Ferling, I., Gazda, M. A., Hoeppner, L. H., Ilangovan, V., Lo, H., Olson, M., Mohamed, S. Y., Sarabipour, S., Varma, A., Walavalkar, K., Wissink, E. M., & Weissgerber, T. L. (2021). Creating clear and informative image-based figures for scientific publications. *PLoS Biology, 19*(3), e3001161. https://doi.org/10.1371/journal.pbio.3001161

Khakshooy, A., Bach, Q., Kasar, V., & Chiappelli, F. (2020). Metascience in bioinformation. *Bioinformation, 16*(1), 4–7. https://doi.org/10.6026/97320630016004

Metascience Symposium. (2019). *Metascience: The Emerging Field of Research on the Scientific Process*. https://www.metascience2019.org/

Metascience Symposium. (2021). *Metascience 2021 Conference*. https://metascience2021.org/

Miles, J. (2022). *Leading Queensland Cancer Researcher Mark Smyth Fabricated Scientific Data, Review Finds*. https://www.abc.net.au/news/2022-01-11/qld-cancer-researcher-mark-smyth-fabricated-data-review-finds/100750208

Pluckrose, H., Lindsay, J., & Boghossian, P. (2021). Understanding the "Grievance studies affair" papers and why they should be reinstated: A response to Geoff Cole. *Sociological Methods & Research, 50*(4), 1916–1936. https://doi.org/10.1177/00491241211009946

Rauh, S., Torgerson, T., Johnson, A. L., Pollard, J., Tritz, D., & Vassar, M. (2020). Reproducible and transparent research practices in published neurology research. *Research Integrity and Peer Review, 5*, 5. https://doi.org/10.1186/s41073-020-0091-5

Roberts, R. G., & PLOS Biology Staff Editors. (2022). The first six years of meta-research at PLOS Biology. *PLoS Biology, 20*(1), e3001553. https://doi.org/10.1371/journal.pbio.3001553

Singal, J. (2019). *Is a Portland Professor Being Railroaded by His University for Criticizing Social-JusticeResearch?*https://nymag.com/intelligencer/2019/01/is-peter-boghossian-getting-railroaded-for-his-hoax.html

Stoeger, T., Gerlach, M., Morimoto, R. I., & Nunes Amaral, L. A. (2018). Large-scale investigation of the reasons why potentially important genes are ignored. *PLoS Biology, 16*(9), e2006643. https://doi.org/10.1371/journal.pbio.2006643

Triunfol, M., & Gouveia, F. C. (2021). What's not in the news headlines or titles of Alzheimer disease articles? #InMice. *PLoS Biology, 19*(6), e3001260. https://doi.org/10.1371/journal.pbio.3001260

Voelkl, B., Vogt, L., Sena, E. S., & Würbel, H. (2018). Reproducibility of preclinical animal research improves with heterogeneity of study samples. *PLoS Biology, 16*(2), e2003693. https://doi.org/10.1371/journal.pbio.2003693

Zeng, X. H., Duch, J., Sales-Pardo, M., Moreira, J. A., Radicchi, F., Ribeiro, H. V., Woodruff, T. K., & Amaral, L. A. (2016). Differences in collaboration patterns across discipline, career stage, and gender. *PLoS Biology, 14*(11), e1002573. https://doi.org/10.1371/journal.pbio.1002573

2 Mitigating biases during the production and dissemination of research

Abstract

A bias is any factor that can undermine the accuracy of a study's conclusions. Biases can occur at any phase of the research process, including during the design of the study to disseminating its findings. In this chapter common biases that can emerge whilst designing, conducting, and disseminating the research are outlined. This chapter's purpose is to assist with the identification and resolution of common biases within research.

Keywords: Artificial Intelligence; Bias; Confirmation Bias; Diversity Badges; Ethnocentric Bias; Gender Bias; Measurement Bias; Place of Publication Bias; Search Engine Bias; Selection Bias; Sex and Gender Equity in Research (SAGER) Guidelines; Time-Lag Bias

Key points

- Biases can distort the accuracy of results and they can occur during the design, collection and analysis of data, and dissemination of the results phases of the study.
- To reduce the occurrence of biases, transparent and comprehensive design, analysis, and dissemination of the study's results are required. For example, the Sex and Gender Equity in Research (SAGER) guidelines can detect and rectify any biases caused by the sample's sex and gender characteristics.
- Artificial intelligence and machine learning can enhance a researcher's ability to collect and analyse studies for a literature review. However, such technological approaches can contain biases that can distort search results. For example, the results obtained from a Google Scholar search are influenced by the geographical location where the search occurred. To rectify this technological limitation, searches of other citation repositories, such as PubMed, are advised.
- Biases can occur within different study types, such as observational studies, clinical trials, laboratory studies, and systematic literature reviews.

2.1 Overview of biases in research

According to the Merriam-Webster (2022) dictionary, a bias is defined as 'systematic error introduced into sampling or testing by selecting or encouraging one outcome or answer over others'. Similarly, according to Šimundić (2013, p. 12) 'bias is any trend or deviation from the truth in data collection, data analysis, interpretation and publication which can cause false conclusions'. As illustrated below, there are different types of biases that can occur during various stages of the research process (see Figure 2.1). Some of these biases are discussed in this chapter.

DOI: 10.4324/9781003510376-2

Figure 2.1 Biases that can occur during the creation of research
Source: Williams et al. (2020, p. 163).

2.2   Biases whilst designing a study

2.2.1   *Ethnocentric bias*

2.2.1.1   *Defining ethnocentric bias*

Ethnocentric bias, which is more colloquially termed 'cultural bias', is the influence that a researcher's cultural background has on their ability to conceive, produce, and then report their study's findings (Brady et al., 2018). The influence of ethnocentric bias on the creation of psychological research has been acknowledged. Multiple scholars who study psychology have claimed that the results of most psychological research published in English are based on samples of participants who live in western, educated, industrialised, rich, and democratic (WEIRD) societies (Brady et al., 2018; Cheon et al., 2020; Kahalon et al., 2021; Masuda et al., 2020; Muthukrishna et al., 2020; Rad et al., 2018). Similarly, within the field of autism spectrum research, it has been acknowledged that there is not much research about people on the autism spectrum from non-WEIRD cultural backgrounds (Jones & Mandell, 2020). Furthermore, scholars who study the autism spectrum assert that the prevailing cultural and ethnic composition within research institutions is a source of ethnocentric bias:

> While many related research professional organizations (Society for Research in Child Development, American Psychological Association, and Society for Neuroscience)

include caucuses to support the needs of Black researchers, no such caucus, committee, or organization currently exists in autism research. INSAR [International Society for Autism Research] has recently published a statement (INSAR Board of Directors and Cultural Diversity Committee, 2020) acknowledging the lack of diversity in autism research and promising to address these issues, but this alone is not sufficient for dismantling the toxic effects of systemic racism in our field. Instead, individuals, organizations, and institutions must work together, taking sustained action to address these inequalities and promote Black autism researchers.

(Jones & Mandell, 2020, p. 1587)

2.2.1.2   *Strategies that can reduce ethnocentric bias*

2.2.1.2.1   MANDATING THE REPORTING OF SAMPLE CHARACTERISTICS

At the moment, most studies report the gender breakdown of their sample but little else. Many fail to disclose the country [the] research took place in, and it seems rare to discuss how wealthy or educated their participants are. We recommend that authors should be required to report a number of other characteristics of their sample, including age, SES [socioeconomic status], ethnicity, religion, and nationality. If this is not possible, authors should acknowledge this and signify that a variable has missing values or data are inapplicable.

(Rad et al., 2018, p. 11403)

As Rad et al. (2018) explained in the quotation above, there is a tendency for scholars to only describe some of the basic characteristics of a sample, such as the age and gender composition of participants. However, the participant's ethnicity and cultural background are rarely documented. In the field of autism spectrum research, for example, Pierce et al. (2014) measured the proportion of studies that contained descriptions of about participant's ethnicity in three autism-specific journals. They reported that the *Journal of Autism and Developmental Disorders* had the most articles that described the participant's ethnicity (36%), followed by *Autism: The International Journal of Research and Practice* (34%), and *Focus on Autism and Developmental Disabilities* (11%). Their finding revealed that most studies published in these autism-specific journals did not report the participant's ethnicity. This lack of reporting means that researchers are unable to measure the extent to which different ethnicities have been studied in research about the autism spectrum. To rectify this situation, journals should mandate that the participant's ethnicity and cultural background are described in the submitted manuscript (Rad et al., 2018).

2.2.1.2.2   CULTURAL DIVERSITY BADGES

To reduce the prospect of cultural bias occurring a cultural diversity badge can be given to manuscripts that contain descriptions about the participant's ethnicity and cultural background. To be awarded such a badge a manuscript's author should declare the cultural diversity of their study's sample and any limitations about applying their results to those from other cultural backgrounds. If the manuscript's author explains such elements, the journal's editor can award a cultural diversity badge (see Figure 2.2) (i.e., More details about Open Science Badges are described in chapter seven). Rad et al. have also proposed that cultural diversity badges should be used to mitigate the occurrence of ethnocentric bias:

Journals are beginning to introduce badges to encourage good methodological research practices. The same should be done to create incentives to sample more diverse populations. To that end, journals could introduce badges to indicate that a manuscript has sampled a population that varies from WEIRD populations on one or more dimensions. A paper that samples a non-Western but educated sample from an industrialized, rich, and democratic society would receive one badge.

(Rad et al., 2018, p. 11404)

Figure 2.2 Hypothetical cultural diversity badge

2.2.1.2.3   DIVERSITY TARGETS

We think it reasonable to suggest the goal of at least 50% of papers sampling populations that deviate from WEIRD populations in at least one dimension. Some may argue that this is low, and that a good goal would be 80%. Setting a clear target is a way of countering implicit biases and current incentive structures. If Psychological Science were to announce that by 2022, half of its papers would include studies sampling at least one non-WEIRD population, it would influence editors, reviewers, and scientists to change their practices to help meet or take advantage of this goal.

(Rad et al., 2018, p. 11404)

As explained in the quotation above, to mitigate the influence of cultural bias Rad et al. have proposed that journals should implement a quota for manuscripts that contain samples of participants from non-WEIRD populations. Such a quota will curtail the production of studies that contain samples of participants from WEIRD populations. If such a quota is not feasible, a publisher can establish a specialised journal in which studies that contain samples of participants from non-WEIRD populations are exclusively published. Alternatively, journals can publish special editions that focus on international research developments from non-WEIRD nations. For example, in 2017, the autism-specific journal *Autism: The International Journal of Research and Practice* published the special issue 'Global Autism Research' (Rice & Lee, 2017).

2.2.1.2.4   INCREASING ETHNIC DIVERSITY ON EDITORIAL BOARDS

There is research about the gender composition of editorial teams for academic journals (Grinnell et al., 2020; Hafeez et al., 2019; Ioannidou & Rosania, 2015; Lobl et al., 2020; Pinho-Gomes et al., 2021; Sougou et al., 2022). In contrast, there is a lack of research about their ethnic composition (Bhaumik & Jagnoor, 2019). It is plausible to assume that this lack of ethnic diversity can inhibit an awareness and appreciation of research about participants from diverse cultural, linguistic, and ethnic backgrounds. However, within some fields, such as autism spectrum research, the lack of ethnic diversity on editorial boards has been acknowledged:

> As Black voices gain more representation within autism research, we hope that they can shape further dialogue on these issues. The journal, Autism, is equally complicit in not addressing these issues. We have far too few Black board members and no Black action editors. We will take a hard look at ourselves and develop concrete steps to improve our anti-racist and inclusive policies. We hope other organizations will do the same.
>
> (Jones & Mandell, 2020, p. 1589)

Arguably, such an acknowledgement can result in greater awareness about the ethnic diversity of editorial teams. However, this acknowledgement alone is insufficient to change the editorial team's ethnic composition. To ensure that editorial teams become more ethnically heterogeneous ethnic diversity quotas should be established and pursued.

2.2.2   Gender bias

2.2.2.1   Defining gender bias

Gender bias occurs when one sex is prioritised in the creation of research instead of the other. With some exceptions, like reproductive research, historically there has been an androcentric (i.e., male-focused) bias within research conducted within WEIRD societies. This bias has resulted in male participants receiving greater attention by researchers than their female counterparts. For example, Salter-Volz et al. (2021) conducted a bibliometric analysis of COVID-19-related case reports that were published in high impact journals from 1 January 2020 to 1 June 2020. Of the 494 case reports examined, 45% (n=221) of patients were male, 30% (n=146) were female, and 25% (n=124) included both sexes.

Within specific fields there is a scant description of the participant's gender and sex which can inhibit the identification and correction of gender bias. For example, reflecting on their results about the extent of the participant's sex and gender in diabetes research being reported Day et al. (2019, p. 7) stated that:

> The results of our analysis demonstrate that sex and gender data are currently poorly reported in diabetes original investigations in the leading general medicine and diabetes-specific journals. Our findings suggest that improvements in sex and gender data reporting are needed order to provide the best possible evidence to inform tailored diabetes care. To enhance the value of research in the field of diabetes and beyond, future efforts should focus on establishing and enforcing compliance with guidelines for sex and gender reporting requirements, particularly in RCTs [randomised controlled trials].

2.2.2.2   Strategies to reduce gender bias

2.2.2.2.1   THE SEX AND GENDER EQUITY IN RESEARCH (SAGER) GUIDELINES

In some disciplines, such as psychology and biomedical research, journals do not provide prospective authors with guidelines that can help them include descriptions about the participant's sex and gender in their manuscript (Cavanaugh & Abu Hussein, 2020; Merriman et al., 2021). Cavanaugh and Abu Hussein (2020), for example, reported that eight (9.0%) of the 89 journals published by the *American Psychological Association* that they examined explicitly gave authors instructions about describing the participant's sex and gender in the title/abstract, introduction, and methods sections in their manuscript. Additionally, they provided no instructions about including the participant's sex and/or gender in the results or discussion sections in their manuscript.

The SAGER guidelines can help improve the prospect of researchers describing the participant's sex and gender in all components of their manuscript (see Appendix 2.1). Several scholars have endorsed these guidelines (Cavanaugh & Abu Hussein, 2020; Heidari et al., 2016). The architects of these guidelines, Heidari et al. (2016, p. 6), have claimed that:

> The SAGER guidelines provide researchers and authors with a tool to standardize sex and gender reporting in scientific publications. They were designed to improve sex and gender reporting of scientific research, serve as a guide for authors and peer-reviewers, be flexible enough to accommodate a wide range of research areas and disciplines and improve the communication of research findings.

The SAGER guidelines can also help journal editors determine if the submitted manuscript contains sufficient details about the participants' sex and gender. Heidari et al. have created a flow diagram to help editors use the SAGER guidelines when evaluating a manuscript. As illustrated in the figure below, when an editor begins the evaluation process, they first decide if sex and gender are relevant to the study. If it is irrelevant, they should confirm that the author has justified their position. Alternatively, if the participant's sex and gender is relevant the editor should examine if this information has been documented in the manuscript's data, design, discussion, and limitations sections (see Figure 2.3).

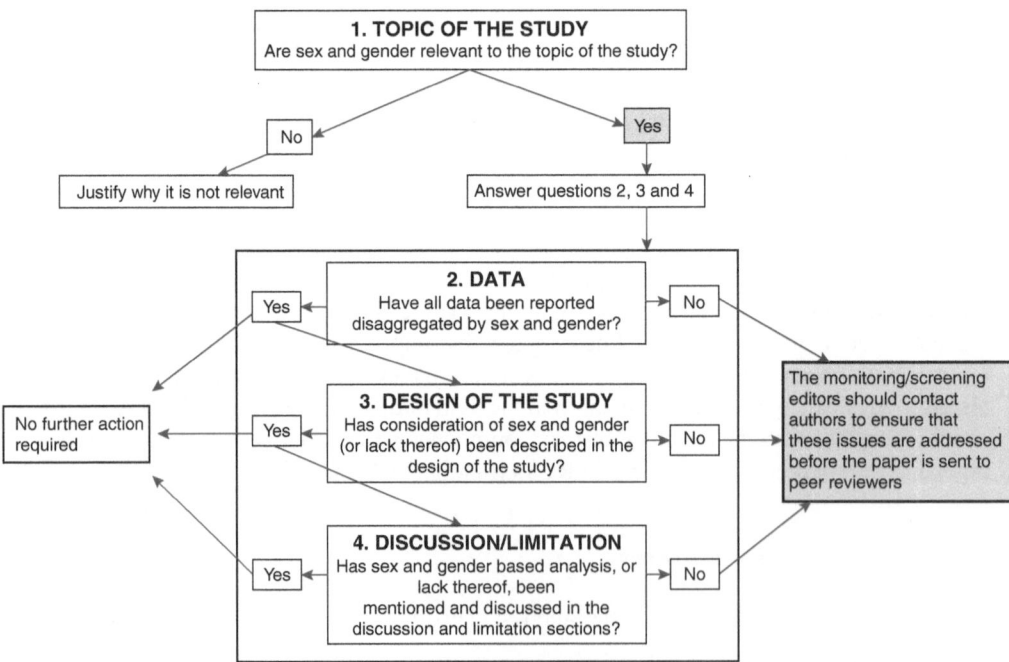

Figure 2.3 Sex and Gender Equity in Research (SAGER) guidelines guiding editors' initial screening of submitted manuscripts

Source: Heidari et al. (2016, p. 6).

The act of describing a participant's sex and gender begins at the proposal stage. There are guidelines that can help researchers incorporate into their research proposals descriptions about the participant's sex and gender characteristics (Hankivsky et al., 2018; Mason, 2020). Hankivsky et al. (2018) have provided a series of sex and gender recommendations for researchers who are creating research proposals (see Appendix 2.2). Additionally, Heidari et al. (2016) have published a checklist for gender-sensitive reporting that can be used to prevent sex and gender bias (see Appendix 2.3).

Despite their utility, journals may be unable to update their instructions to authors or electronic submission systems to comply with the SAGER criteria. Large publishers typically provide authors with similar instructions for all their journals and seldom allow authors to change their standards. A push from journal editors might encourage publishers to include the SAGER criteria along with other reporting rules across all their journals. Furthermore, submission systems may be modified to include a checkbox for authors to indicate that their manuscript adheres to the SAGER guidelines. Additionally, questions about if the submitted manuscript complies with the SAGER guidelines can be added to the evaluation form that peer reviewers complete when they are evaluating a submitted manuscript (Peters et al., 2021).

There are also issues regarding the additional strain put on peer reviewers by applying the SAGER criteria. While verifying for sex and gender reporting takes time, the workload on peer reviewers can be reduced if submitted publications are vetted before peer review. In most journals, screening papers for conformity to editorial criteria is already a standard procedure (Peters et al., 2021).

Journal editors may lack the time, talent, and resources to implement or enforce the SAGER guidelines as a formal policy. Journals with few issues per year may find it especially difficult, given occasional board meetings and conflicting objectives. Nonetheless, as a means of improving scientific quality, journals should explore adopting the SAGER guidelines and other reporting requirements. Support from publishers might make it easier to implement the SAGER guidelines across all publications (Peters et al., 2021).

2.2.2.2.2   GENDER DIVERSITY BADGES

As explained previously, a cultural diversity badge might improve the inclusion of participants from culturally and ethnically diverse backgrounds in research. It is feasible that the same approach can be used to enhance the inclusion of participants with different sex and gender characteristics in research. To earn a hypothetical 'sex and gender diversity' badge, the author should explain how their study complies with the SAGER guidelines (see Figure 2.4).

Figure 2.4 Hypothetical sex and gender diversity badge

2.3   Biases during data collection and analysis

2.3.1   *Cognitive bias*

> The premise that best explains how cognitive biases may undermine the interpretation of a phenomenon is the fallibility of the human brain. Indeed, we are cognitively predisposed to interpret facts based on a number of fallible systems … Accordingly we tend to draw conclusions or to find a quick solution to a given problem even though we lack all the information that is required to do so. Although this attitude is clearly advantageous in some life situations, it is a limiting factor when novel research findings confute our beliefs.
>
> (Puzzo & Conti, 2021, p. 2)

As explained in the quote above our cognitive processes, in particular our inclination to not expend much intellectual effort, can inhibit our ability to interpret data and accurately formulate conclusions. Expressed differently, researchers are humans and sometimes their neurological characteristics can undermine their interpretation of information and prevent them from creating conclusions that are truthful (Puzzo & Conti, 2021). Wakefield et al.'s (1998) study which concluded that children who received the Measles, Mumps, and Rubella (MMR) vaccine were more likely to develop autism is an example of cognitive bias. There is an abundance of literature that has proven that the MMR vaccine does not contribute to children developing autism (Dales et al., 2001; DeStefano et al., 2004; Honda et al., 2005; Kaye et al., 2001). Due to this evidence, it is plausible to assume that children receiving their MMR vaccine and then exhibiting their initial characteristics of autism are just coincidental events (Bennett et al., 2018). Despite being coincidental some still confuse coincidental events as causational events.

2.3.2   *Confirmation bias*

Confirmation bias is a tendency to preferentially favour information that is consistent and/or reinforces one's opinions or hypothesis. To explain this concept Suzuki and Yamamoto (2021, pp. 1–2) provided the following example about how confirmation bias can distort the information someone examines from the internet about different foods:

> … assume that user X, who is health conscious, learns on TV that food Y, which uses genetic modification, is harmful to health and distrusts food Y. When user X performs a web search to obtain information about food Y's safety, they unconsciously seek to support the idea that food Y is harmful to their health; therefore, user X will preferentially browse negative information about food Y, even if that information is incorrect or low-quality. Thus, confirmation bias can be a significant problem in web search behavior because confirmation bias that occurs when users search the web for information about food, clothing, housing, and politics can significantly impact society.

The influence that confirmation bias has on a researcher's ability to conduct objective and impartial research has been examined. For example, for Meppelink et al.'s (2019) study 480 parents raising young children (i.e., aged 0–4 years) were required to complete three tasks. First, they completed an online survey about their beliefs about vaccinations. Second, from a list of vaccine-related messages they were asked to select what messages they found the most appealing. Third, they had to evaluate the credibility, usefulness, and convincingness of two texts that described vaccination either positively or negatively. They discovered that parents in their study selectively exposed themselves to inaccurate information that was consistent with their beliefs about childhood vaccines instead of truthful information that actually challenged or contradicted their personal beliefs. Additionally, when they encountered false information that matched their beliefs, they were more likely to deem it credible, useful, and convincing.

2.3.3   *Selection bias*

Selection bias can occur during the design phase of the study, especially when drafting the study's inclusion criteria and/or when participants are recruited (Smith & Noble, 2014). To illustrate the occurrence of selection bias during the study's recruitment phase Coggon et al. (2009) provided the following example:

> Suppose that an investigator wishes to estimate the prevalence of heavy alcohol consumption (more than 21 units a week) in adult residents of a city. He might try to do this by selecting a random sample from all the adults registered with local general practitioners, and sending them a postal questionnaire about their drinking habits. With this design, one source of error would be the exclusion from the study sample of those residents not registered with a doctor. These excluded subjects might have different patterns of drinking from those included in the study. Also, not all of the subjects selected for study will necessarily complete and return questionnaires, and non-responders may have different drinking habits from those who take the trouble to reply. Both of these deficiencies are potential sources of selection bias.

Using a 'results free' manuscript submission process can assist in the identification and prevention of selection bias. This submission process has two phases. First, the author submits a research plan that explains the objective of the study, the number of participants that will be recruited, how the participants will be recruited, and how the collected data will be analysed. In this research plan strategies that can mitigate the occurrence of selection bias are explained. If peer reviewers deem the research plan to be methodologically robust, including strategies to reduce selection bias, the author is then permitted to conduct the study. Alternatively, if the plan is judged to be insufficient then it is either rejected or the author is invited to make modifications. The second phase involves the author conducting the study and then submitting a complete manuscript for review. During this second phase peer reviewers examine the complete study and decide if it is of suitable quality to be published, including if the study has reduced any instances of selection bias.

2.3.4   *Data collection bias*

Data collection bias can occur when the researcher's personal beliefs, including their personal beliefs about the topic they are examining or the participants, can influence the way that they collect data (Smith & Noble, 2014). To reduce the impact of this bias a researcher can engage in a reflexive exercise and/or other research activities to safeguard data integrity (Korstjens & Moser, 2018) (see Tables 2.1 and 2.2).

2.3.5   *Measurement bias*

Measurement bias is the misapplication of a measurement instrument that results in inaccurate findings being reported. It can occur in qualitative, quantitative, and retrospective studies. In qualitative research there are three ways that measurement bias can occur. First, when an inappropriate measurement tool is used. For example, measuring a patient's decision-making capabilities with a patient satisfaction survey. Second, when a measurement instrument is misapplied to an inappropriate group. For example, measuring a child's pain and discomfort using an instrument designed to only measure an adult's pain and discomfort. Third, when a measurement instrument has not

Table 2.1 Definitions of quality criteria in qualitative research

Characteristic	Definition
Credibility	The confidence that can be placed in the truth of the research findings. Credibility establishes whether the research findings represent plausible information drawn from the participants' original data and is a correct interpretation of the participants' original views.
Transferability	The degree to which the results of qualitative research can be transferred to other contexts or settings with other respondents. The researcher facilitates the transferability judgement by a potential user through thick description.
Dependability	The stability of findings over time. Dependability involves participants' evaluation of the findings, interpretation and recommendations of the study such that all are supported by the data as received from participants of the study.
Confirmability	The degree to which the findings of the research study could be confirmed by other researchers. Confirmability is concerned with establishing that data and interpretations of the findings are not figments of the inquirer's imagination, but clearly derived from the data.
Reflexivity	The process of critical self-reflection about oneself as researcher (own biases, preferences, preconceptions), and the research relationship (relationship to the respondent, and how the relationship affects participant's answers to questions).

Source: Korstjens and Moser (2018, p. 121).

been properly calibrated. For example, a digital thermometer might not have been reset to zero degrees Celsius before being used. In retrospective studies, measurement bias can occur when a participant does not accurately recall events. For example, some participants might have 'rose coloured glasses' and recall positive experiences when they were in fact abused or mistreated. In qualitative research, measurement bias can happen when participants are asked closed questions. For example, when a patient is asked 'Do you find the health service poor?' they are more inclined to respond with either a 'yes' or a 'no'. To gain a more comprehensive response a suitable interview question could be 'Please describe your last visit to hospital?' (Smith & Noble, 2014).

2.3.6   *Search engine bias*

The results that are obtained from an internet search engine are influenced by the search engine's algorithmic design, when and where the search was conducted, and the user's licensing agreement. Pozsgai and colleagues (2021) evaluated if the geographical location where the internet search occurred had an impact on the consistency of search results obtained. To test this phenomenon, they conducted time-synchronised searches using identical search terms and search engines but at different international institutional locations. They concluded that the location where the search occurred did influence the number and specific articles retrieved using Google Scholar and Web of Science's Core Collection. Based on these results, Pozsgai and colleagues recommended seven strategies to ensure that search results from the internet are robust and reproducible (see Appendix 2.4).

2.4   Biases during the reporting and dissemination of research

2.4.1   *Time-lag bias*

Time-lag bias occurs when studies with positive results are published faster than those with negative results. For example, a study that shows that drug X can treat lung cancer is published quicker than a study that shows that drug Y cannot treat lung cancer (Reyes et al., 2011).

Table 2.2 Strategies to ensure trustworthiness in qualitative research

Criteria	Strategy	Definition
Credibility	Prolonged engagement	Lasting presence during observation of long interviews or long-lasting engagement in the field with participants. Investing sufficient time to become familiar with the setting and context, to test for misinformation, to build trust, and to get to know the data to get rich data.
	Persistent observation	Identifying those characteristics and elements that are most relevant to the problem or issue under study, on which you will focus in detail.
	Triangulation	Using different data sources, investigators and methods of data collection. • *Data triangulation* refers to using multiple data sources in time (gathering data in different times of the day or at different times in a year), space (collecting data on the same phenomenon in multiple sites or test for cross-site consistency) and person (gathering data from different types or level of people, e.g., individuals, their family members, and clinicians). • *Investigator triangulation* is concerned with using two ore researchers to make coding, analysis and interpretation decisions. • *Method triangulation* means using multiple methods of data collection.
	Member check	Feeding back data, analytical categories, interpretations and conclusions to members of those groups from whom the data were originally obtained. It strengthens the data, especially because researcher and respondents look at the data with different eyes.
Transferability	Thick description	Describing not just the behaviour and experiences, but their context as well, so that the behaviour and experiences become meaningful to an outsider.
Dependability and confirmability	Audit trail	Transparently describing the research steps taken from the start of a research project to the development and reporting of the findings. The records of the research path are kept throughout the study.
Reflexivity	Diary	Examining one's own conceptual lens, explicit and implicit assumptions, preconceptions and values, and how these affect research decisions in all phases of qualitative studies.

Source: Korstjens and Moser (2018, p. 121).

Time-lag bias can undermine the reliability of research because promising results often receive inadequate scrutiny because they are published quickly. This point has been explained by Hopewell and colleagues (2007, p. 3), who stated:

> If time-lag bias exists, such that the results of studies with more striking findings become available (through publication) sooner that those with less striking findings, this could be harmful. If the time taken to publish trials with positive results is years shorter than it is for those with negative or null results, new interventions will be accepted as effective in the absence of evidence to the contrary, even though that evidence may already have been gathered. Time-lag bias might also introduce bias into systematic reviews if a study's inclusion is related to the timing of the availability of its data.

To reduce any impacts caused by time-lag bias journals that only publish studies with negative results can be established. Such journals can ensure that studies with negative results are published just as those that report positive results.

2.4.2   *Place of publication bias*

Place of publication bias happens when a study is more inclined to be cited by others because it has been published in a journal that is perceived to be prestigious (Williams et al., 2020). This bias can cause credible studies that contain important results being overlooked because they happen to have been published in less prestigious journals. To mitigate this bias, a researcher should search multiple databases with multiple keyword permutations to ensure that the largest number of relevant studies possible are identified and included in their study or research activities (Bramer et al., 2017; Justesen et al., 2021).

2.4.3   *Citation bias*

The selection of studies based on their findings is referred to as 'citation bias'. Duyx and colleagues (2017) gathered and quantified published research about citation bias across different scientific disciplines. They categorised 52 studies by their scientific field, selection method, and other variables. To compare the impact of favourable vs. unfavourable outcomes on subsequent citations, they conducted a random-effects meta-analysis. Finally, they examined other factors responsible for citation bias that was documented in the literature about citation bias. They concluded that articles about biological sciences showed the greatest evidence of citation bias while articles about natural sciences showed the least. In comparison to studies with nonsignificant results, those with statistically significant results received 1.6 (95% confidence interval [CI] = 1.3–1.8) times as many citations. Studies that clearly stated that they had uncovered evidence to support their hypothesis were referenced 2.7 (95% CI = 2.0–3.7) times more frequently. More often than any other examined factor, the results in the article and the journal's impact factor influenced it's citation rate. The occasional act of citation bias may have a negligible impact, but its accumulative effect is an overrepresentation of favourable findings in the scientific literature. Consequently, scientific progress is impeded, incorrect conclusions and choices are made, and eventually the reputation of science is tarnished.

2.4.4   *Checklists to detect bias in manuscripts*

There are several checklists, such as those published by JBI and the use the *Enhancing the QUAlity and Transparency Of health Research* (EQUATOR) checklists, that authors and editors can use to improve the quality of research and identify any biases that are embedded within the study. These checklists will be discussed in more detail in the next chapter.

2.5   Biases in artificial intelligence

2.5.1   *The role of artificial intelligence in the creation of literature reviews*

Some researchers are using artificial intelligence (AI) programs to augment and improve their creation of research. However, for brevity, the role and potential biases that AI generate in the creation of systematic literature reviews are discussed in this section (Hamel et al., 2021). Chai and colleagues (2021) have outlined the relationship between researchers and AI programs. They propose that researchers identify articles that they believe would be relevant to the systematic literature review. They then select and give to the AI program one or two 'seed' articles. The AI program then removes any duplicate records. Afterwards, a machine learning cycle begins, in which the AI program retrieves and then ranks potentially relevant articles for the researcher to evaluate. Articles that the researcher deems as suitable are then given back to the AI program for other similar articles to be retrieved. Once this cycle is complete multiple researchers discuss and resolve any conflicts or uncertainties in the sample of articles that the AI program has suggested for the systematic literature review. Once any discrepancies are resolved all articles are exported for full-text review (see Figure 2.5) (Chai et al., 2021).

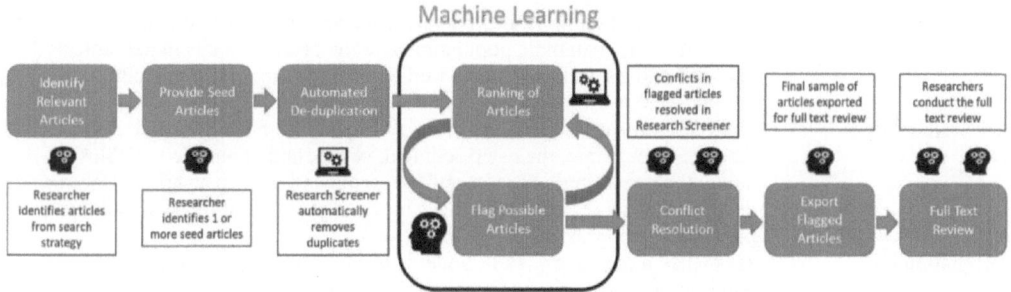

Figure 2.5 Artificial intelligence and research screener assisted screening process

Source: Chai et al. (2021, p. 4).

Note: Head symbol = elements that require human input; Computer symbol = elements augmented by the artificial intelligence program.

2.5.2   *Limitations of AI*

> Since computers are psychologically associated with pure impartial logic, it was often thought that computers could be an ideal solution to the human problem of discrimination and prejudice in complex decisions. We now know, however, that machine learning algorithms and the data used to train them are generally a product of human design and can also be flawed.
>
> (Fletcher et al., 2021, p. 6)

As explained by Fletcher and colleagues, since AI programs are designed by humans they can contain human imperfections and prejudices. Norori and colleagues (2021) described three biases that might impede AI algorithms from generating accurate conclusions. These three biases are data-driven bias, algorithmic bias, and human bias. Data-driven bias occurs when the AI program generates results based on a dataset that has missing information and/or biased samples. For example, AI programs might retrieve studies for a literature review about people on the autism spectrum and support programs. However, the results will have a gender bias in favour of males since most studies about this topic contain samples of males on the autism spectrum (Edwards et al., 2012; Watkins et al., 2014). Alternatively, an algorithm may be unable to identify skin cancer in non-Caucasian patients because the algorithm has been trained using a dataset of images of Caucasian people with skin cancer (Norori et al., 2021). Algorithmic bias occurs when the algorithm's design creates biased outcomes. For example, the algorithm that produces the search results for Google Scholar creates different search results depending on the geographical region where the search occurred and the user's licence agreement (Pozsgai et al., 2021). Human bias is any bias in an AI program that is based on human prejudices. This bias can be difficult to detect and resolve

Table 2.3 Source of undesirable bias in artificial intelligence with examples in health research and practice

Source of bias in artificial intelligence	Description
Historical bias	Arises even if the data is perfectly measured and sampled, when the world as it is leads a model to produce outcomes that are not desired. For example, incorrectly assuming that Human Immunodeficiency Virus is inherently linked to homosexual and bisexual men as its prevalence is higher in this population.
Representation bias	Occurs when certain parts of the input space are underrepresented. For example, European male populations are the primary focus in genomics research and its derived clinical findings, neglecting other ethnicities and populations.
Measurement bias	Occurs when measured data are often proxies for some ideal features and labels. For example, the use of clinical, social, and cognitive variables to detect the prodromal phase in schizophrenia and other psychotic disorders despite of observed sex differences in the expression of those symptoms and their associated risk for psychosis.
Aggregation bias	Arises when a one-size-fits-all model is used for groups with different conditional distributions. For example, for the diagnosis and monitoring of diabetes, haemoglobin A1c (HbA1c) levels are routinely used, despite of differences associated with ethnicities and gender.
Evaluation bias	Occurs when the evaluation and/or benchmark data for an algorithm does not represent the target population. For example, underperformance of commercial facial recognition algorithm in dark-skinned female faces as most benchmark face image datasets come from white men.
Algorithmic Bias	Occurs when bias is introduced in the algorithm consciously or unconsciously in ad-hoc solutions. For example, by using health care cost as a proxy feature for health status without correcting for existing inequalities in health access, a commercial algorithm to predict health care needs was found to exhibit significant racial discrimination.

Source: Cirillo et al. (2020, p. 5).

because it 'can result from long-held societal prejudices that may be subtle at the level of society, and amplified by AI and large datasets' (Norori et al., 2021, p. 4). Cirillo and colleagues (2020) have also defined biases that are within AI programs (see Table 2.3).

Norori and colleagues have outlined a series of Open Science principles that are intended to identify and remove any biases that are inherent in the creation and usage of AI programs. These principles include incorporating into a dataset the characteristics of minority groups so that the algorithm can make more accurate decisions and publishing the algorithm code so that others can review its design and identify any flaws or biases. With the intention of applying Open Science principles to the development of AI programs, Fletcher and colleagues (2021) have proposed a series of recommendations (see Table 2.4).

2.6 Conclusion

This chapter has explained some common biases that can occur whilst designing, conducting, and reporting the results of a study. To accomplish this goal, this chapter began with a broad explanation of bias and explanations about several biases in the design, conduct, and reporting of a study's results. It is expected that this chapter will give the reader the skills needed to recognise and minimise biases in their study.

Table 2.4 Recommendations by Fletcher and colleagues to improve artificial intelligence programs used in global health research

Characteristic	Explanation
Question appropriate use	Machine learning models are designed to answer specific questions. However, since many health care decisions can have unintended consequences, it is always important to review if we are posing the right question. As the use of machine learning and AI is extended beyond diagnostic tools into questions related to health care access, medical triage, and insurance coverage, it is also important to remember that no decision-making tool is perfect, and that certain important decisions should perhaps be reserved for humans and not machines.
Maintain transparency for critical decisions	As the complexity of machine learning models continues to increase, and new models are invented (e.g., deep neural nets), the ability to explain the decision of a computer may become increasingly challenging. The use of "black box" models is useful in certain cases but not others. For decisions that involve human input, such as patient diagnosis, it is recommended to use interpretable models that will enable review and consensus from human staff.
Enforce transparency in data and algorithms	Algorithms are not universal, and are only valid when used properly. Algorithms are critically dependent on the specific training data that was used for development, as well as the optimisation criteria used to tune the model. In order to avoid problems with fairness, it is important that this information be disclosed to organisations and patients that will be using the model. The government regulation of algorithms is in a very early stage, but it is perhaps inevitable that some type of regulation will be put in place, in a manner analogous to the FDA regulation of pharmaceuticals; "off-label" uses of an algorithm can produce unexpected or unfair results and should be avoided.
Address and Respect Bias	Bias should be examined at each level of the computation, and the individual features used in the training data should themselves be examined for bias. While the bias found in other application domains of machine learning (finance, employment, law enforcement, etc.) may often be due to sampling bias or implicit cultural bias, the domain of health also contains true systematic bias inherent in biological processes which may not be possible to mitigate or "repair". In the domain of health, there are true genetic differences across different races and ethnic groups which affect disease prevalence, and these differences cannot (and should not) be ignored. If it is revealed that a particular algorithm consistently produces very different results for one patient group vs. another, it is generally best to design a separate algorithm for each group rather than try to create a universal algorithm that will likely perform poorly on both groups.
Agree on a Fairness Metric	Since Fairness can be defined in multiple ways, it is important in each case, to decide which fairness metric will be applied and agree on the criteria that are being optimised. It is important to recognise that this step involves much more than technical expertise, and requires the participation of all stakeholders, including the individual groups that may be impacted by the use of the algorithm. While individual fairness is a good ideal, most laws are written with respect to group fairness, and thus when machine learning decisions are applied across multiple groups, it should be recognised that trade-offs and compromises will often need to be made, to reconcile how the benefit and the risk will be shared across all groups.

Source: Fletcher et al. (2021).

Additional readings

DeCamp, M., & Lindvall, C. (2020). Latent bias and the implementation of artificial intelligence in medicine. *Journal of the American Medical Informatics Association: JAMIA, 27*(12), 2020–2023. https://doi.org/10.1093/jamia/ocaa094

Muthukrishna, M., Bell, A. V., Henrich, J., Curtin, C. M., Gedranovich, A., McInerney, J., & Thue, B. (2020). Beyond Western, educated, industrial, rich, and democratic (WEIRD) psychology: Measuring and mapping scales of cultural and psychological distance. *Psychological Science, 31*(6), 678–701. https://doi.org/10.1177/0956797620916782

Vokinger, K. N., Feuerriegel, S., & Kesselheim, A. S. (2021). Mitigating bias in machine learning for medicine. *Communications Medicine, 1*, 25. https://doi.org/10.1038/s43856-021-00028-w

References

Bennett, M., Webster, A. A., Goodall, E., & Rowland, S. (2018). Challenging the public's perception of life on autism spectrum: The impact of the vaccination myth. In *Life on the Autism Spectrum* (pp. 37–60). Springer, Singapore. https://doi.org/10.1007/978-981-13-3359-0_3

Bhaumik, S., & Jagnoor, J. (2019). Diversity in the editorial boards of global health journals. *BMJ Global Health, 4*(5), e001909. https://doi.org/10.1136/bmjgh-2019-001909

Brady, L. M., Fryberg, S. A., & Shoda, Y. (2018). Expanding the interpretive power of psychological science by attending to culture. *Proceedings of the National Academy of Sciences, 115*(45), 11406–11413. https://doi.org/10.1073/pnas.1803526115

Bramer, W. M., Rethlefsen, M. L., Kleijnen, J., & Franco, O. H. (2017). Optimal database combinations for literature searches in systematic reviews: A prospective exploratory study. *Systematic Reviews, 6*(1), 245. https://doi.org/10.1186/s13643-017-0644-y

Cavanaugh, C., & Abu Hussein, Y. (2020). Do journals instruct authors to address sex and gender in psychological science? *Research Integrity and Peer Review, 5*, 14. https://doi.org/10.1186/s41073-020-00100-4

Chai, K., Lines, R., Gucciardi, D. F., & Ng, L. (2021). Research screener: A machine learning tool to semi-automate abstract screening for systematic reviews. *Systematic Reviews, 10*(1), 93. https://doi.org/10.1186/s13643-021-01635-3

Cheon, B. K., Melani, I., & Hong, Y. Y. (2020). How USA-centric is psychology? An archival study of implicit assumptions of generalizability of findings to human nature based on origins of study samples. *Social Psychological and Personality Science, 11*(7), 928–937. https://doi.org/10.1177/1948550620927269

Cirillo, D., Catuara-Solarz, S., Morey, C., Guney, E., Subirats, L., Mellino, S., Gigante, A., Valencia, A., Rementeria, M. J., Chadha, A. S., & Mavridis, N. (2020). Sex and gender differences and biases in artificial intelligence for biomedicine and healthcare. *NPJ Digital Medicine, 3*, 81. https://doi.org/10.1038/s41746-020-0288-5

Coggon, D., Barker, D., & Rose, G. (2009). *Epidemiology for the Uninitiated, London.* John Wiley & Sons, UK.

Dales, L., Hammer, S. J., & Smith, N. J. (2001). Time trends in autism and in MMR immunization coverage in California. *JAMA, 285*(9), 1183–1185. https://doi.org/10.1001/jama.285.9.1183

Day, S., Wu, W., Mason, R., & Rochon, P. A. (2019). Measuring the data gap: Inclusion of sex and gender reporting in diabetes research. *Research Integrity and Peer Review, 4*(1), 1–8. https://doi.org/10.1186/s41073-019-0068-4

DeStefano, F., Bhasin, T. K., Thompson, W. W., Yeargin-Allsopp, M., & Boyle, C. (2004). Age at first measles-mumps-rubella vaccination in children with autism and school-matched control subjects: A population-based study in metropolitan Atlanta. *Pediatrics, 113*(2), 259–266. https://doi.org/10.1542/peds.113.2.259

Duyx, B., Urlings, M., Swaen, G., Bouter, L. M., & Zeegers, M. P. (2017). Scientific citations favor positive results: A systematic review and meta-analysis. *Journal of Clinical Epidemiology, 88*, 92–101. https://doi.org/10.1016/j.jclinepi.2017.06.002

Edwards, T. L., Watkins, E. E., Lotfizadeh, A. D., & Poling, A. (2012). Intervention research to benefit people with autism: How old are the participants? *Research in Autism Spectrum Disorders, 6*(3), 996–999. https://doi.org/10.1016/j.rasd.2011.11.002

Fletcher, R. R., Nakeshimana, A., & Olubeko, O. (2021). Addressing fairness, bias, and appropriate use of artificial intelligence and machine learning in global health. *Frontiers in Artificial Intelligence, 3*, 561802. https://doi.org/10.3389/frai.2020.561802

Grinnell, M., Higgins, S., Yost, K., Ochuba, O., Lobl, M., Grimes, P., & Wysong, A. (2020). The proportion of male and female editors in women's health journals: A critical analysis and review of the sex gap. *International Journal of Women's Dermatology, 6*(1), 7–12. https://doi.org/10.1016/j.ijwd.2019.11.005

Hafeez, D. M., Waqas, A., Majeed, S., Naveed, S., Afzal, K. I., Aftab, Z., Zeshan, M., & Khosa, F. (2019). Gender distribution in psychiatry journals' editorial boards worldwide. *Comprehensive Psychiatry, 94*, 152119. https://doi.org/10.1016/j.comppsych.2019.152119

Hamel, C., Hersi, M., Kelly, S. E., Tricco, A. C., Straus, S., Wells, G., Pham, B., & Hutton, B. (2021). Guidance for using artificial intelligence for title and abstract screening while conducting knowledge syntheses. *BMC Medical Research Methodology, 21*(1), 285. https://doi.org/10.1186/s12874-021-01451-2

Hankivsky, O., Springer, K. W., & Hunting, G. (2018). Beyond sex and gender difference in funding and reporting of health research. *Research Integrity and Peer Review, 3*, 6. https://doi.org/10.1186/s41073-018-0050-6

Heidari, S., Babor, T. F., De Castro, P., Tort, S., & Curno, M. (2016). Sex and gender equity in research: Rationale for the SAGER guidelines and recommended use. *Research Integrity and Peer Review, 1*, 2. https://doi.org/10.1186/s41073-016-0007-6

Honda, H., Shimizu, Y., & Rutter, M. (2005). No effect of MMR withdrawal on the incidence of autism: A total population study. *Journal of Child Psychology and Psychiatry, and Allied Disciplines, 46*(6), 572–579. https://doi.org/10.1111/j.1469-7610.2005.01425.x

Hopewell, S., Clarke, M., Stewart, L., & Tierney, J. (2007). Time to publication for results of clinical trials. *The Cochrane Database of Systematic Reviews, 2007*(2), MR000011. https://doi.org/10.1002/14651858.MR000011.pub2

Ioannidou, E., & Rosania, A. (2015). Under-representation of women on dental journal editorial boards. *PloS One, 10*(1), e0116630. https://doi.org/10.1371/journal.pone.0116630

Jones, D. R., & Mandell, D. S. (2020). To address racial disparities in autism research, we must think globally, act locally. *Autism: The International Journal of Research and Practice, 24*(7), 1587–1589. https://doi.org/10.1177/1362361320948313

Justesen, T., Freyberg, J., & Schultz, A. (2021). Database selection and data gathering methods in systematic reviews of qualitative research regarding diabetes mellitus – An explorative study. *BMC Medical Research Methodology, 21*(1), 94. https://doi.org/10.1186/s12874-021-01281-2

Kahalon, R., Klein, V., Ksenofontov, I., Ullrich, J., & Wright, S. C. (2021). Mentioning the sample's country in the article's title leads to bias in research evaluation. *Social Psychological and Personality Science*, 19485506211024036. https://doi.org/10.1177/19485506211024036

Kaye, J. A., del Mar Melero-Montes, M., & Jick, H. (2001). Mumps, measles, and rubella vaccine and the incidence of autism recorded by general practitioners: A time trend analysis. *BMJ (Clinical research ed.), 322*(7284), 460–463. https://doi.org/10.1136/bmj.322.7284.460

Korstjens, I., & Moser, A. (2018). Series: Practical guidance to qualitative research. Part 4: Trustworthiness and publishing. *The European Journal of General Practice, 24*(1), 120–124. https://doi.org/10.1080/13814788.2017.1375092

Lobl, M., Grinnell, M., Higgins, S., Yost, K., Grimes, P., & Wysong, A. (2020). Representation of women as editors in dermatology journals: A comprehensive review. *International Journal of Women's Dermatology, 6*(1), 20–24. https://doi.org/10.1016/j.ijwd.2019.09.002

Mason, R. (2020). Doing better: Eleven ways to improve the integration of sex and gender in health research proposals. *Research Integrity and Peer Review, 5*(1), 15. https://doi.org/10.1186/s41073-020-00102-2

Masuda, T., Batdorj, B., & Senzaki, S. (2020). Culture and attention: Future directions to expand research beyond the geographical regions of WEIRD cultures. *Frontiers in Psychology, 11*, 1394. https://doi.org/10.3389/fpsyg.2020.01394

Meppelink, C. S., Smit, E. G., Fransen, M. L., & Diviani, N. (2019). "I was Right about Vaccination": Confirmation bias and health literacy in online health information seeking. *Journal of Health Communication, 24*(2), 129–140. https://doi.org/10.1080/10810730.2019.1583701

Merriman, R., Galizia, I., Tanaka, S., Sheffel, A., Buse, K., & Hawkes, S. (2021). The gender and geography of publishing: A review of sex/gender reporting and author representation in leading general medical and global health journals. *BMJ Global Health, 6*(5), e005672. https://doi.org/10.1136/bmjgh-2021-005672

Merriman Webster. (2022). *Bias*. https://www.merriam-webster.com/dictionary/bias

Muthukrishna, M., Bell, A. V., Henrich, J., Curtin, C. M., Gedranovich, A., McInerney, J., & Thue, B. (2020). Beyond western, educated, industrial, rich, and democratic (WEIRD) psychology: Measuring and mapping scales of cultural and psychological distance. *Psychological Science, 31*(6), 678–701. https://doi.org/10.1177/0956797620916782

Norori, N., Hu, Q., Aellen, F. M., Faraci, F. D., & Tzovara, A. (2021). Addressing bias in big data and AI for health care: A call for open science. *Patterns (New York, N.Y.), 2*(10), 100347. https://doi.org/10.1016/j.patter.2021.100347

Peters, S., Babor, T. F., Norton, R. N., Clayton, J. A., Ovseiko, P. V., Tannenbaum, C., & Heidari, S. (2021). Fifth anniversary of the Sex and Gender Equity in Research (SAGER) guidelines: Taking stock and looking ahead. *BMJ Global Health, 6*(11), e007853. https://doi.org/10.1136/bmjgh-2021-007853

Pierce, N. P., O'Reilly, M. F., Sorrells, A. M., Fragale, C. L., White, P. J., Aguilar, J. M., & Cole, H. A. (2014). Ethnicity reporting practices for empirical research in three autism-related journals. *Journal of Autism and Developmental Disorders, 44*(7), 1507–1519. https://doi.org/10.1007/s10803-014-2041-x

Pinho-Gomes, A. C., Vassallo, A., Thompson, K., Womersley, K., Norton, R., & Woodward, M. (2021). Representation of women among editors in chief of leading medical journals. *JAMA Network Open, 4*(9), e2123026. https://doi.org/10.1001/jamanetworkopen.2021.23026

Pozsgai, G., Lövei, G. L., Vasseur, L., Gurr, G., Batáry, P., Korponai, J., Littlewood, N. A., Liu, J., Móra, A., Obrycki, J., Reynolds, O., Stockan, J. A., VanVolkenburg, H., Zhang, J., Zhou, W., & You, M. (2021). Irreproducibility in searches of scientific literature: A comparative analysis. *Ecology and Evolution, 11*(21), 14658–14668. https://doi.org/10.1002/ece3.8154

Puzzo, D., & Conti, F. (2021). Conceptual and methodological pitfalls in experimental studies: An overview, and the case of Alzheimer's disease. *Frontiers in Molecular Neuroscience, 14*, 684977. https://doi.org/10.3389/fnmol.2021.684977

Rad, M. S., Martingano, A. J., & Ginges, J. (2018). Toward a psychology of Homo sapiens: Making psychological science more representative of the human population. *Proceedings of the National Academy of Sciences, 115*(45), 11401–11405. https://doi.org/10.1073/pnas.1721165115

Reyes, M. M., Panza, K. E., Martin, A., & Bloch, M. H. (2011). Time-lag bias in trials of pediatric antidepressants: A systematic review and meta-analysis. *Journal of the American Academy of Child and Adolescent Psychiatry, 50*(1), 63–72. https://doi.org/10.1016/j.jaac.2010.10.008

Rice, C. E., & Lee, L. C. (2017). Expanding the global reach of research in autism. *Autism: The International Journal of Research and Practice, 21*(5), 515–517. https://doi.org/10.1177/1362361317704603

Salter-Volz, A. E., Oyasu, A., Yeh, C., Muhammad, L. N., & Woitowich, N. C. (2021). Sex and gender bias in covid-19 clinical case reports. *Frontiers in Global Women's Health, 2*. https://doi.org/10.3389/fgwh.2021.774033

Šimundić, A. M. (2013). Bias in research. *Biochemia Medica, 23*(1), 12–15. http://dx.doi.org/10.11613/BM.2013.003

Smith, J., & Noble, H. (2014). Bias in research. *Evidence-Based Nursing, 17*(4), 100–101. https://doi.org/10.1136/eb-2014-101946

Sougou, N. M., Ndiaye, O., Nabil, F., Folayan, M. O., Sarr, S. C., Mbaye, E. M., & Martínez-Pérez, G. Z. (2022). Barriers of West African women scientists in their research and academic careers: A qualitative research. *PloS One, 17*(3), e0265413. https://doi.org/10.1371/journal.pone.0265413

Suzuki, M., & Yamamoto, Y. (2021). Characterizing the influence of confirmation bias on web search behavior. *Frontiers in Psychology, 12*, 771948. https://doi.org/10.3389/fpsyg.2021.771948

Wakefield, A. J., Murch, S. H., Anthony, A., Linnell, J., Casson, D. M., Malik, M., Berelowitz, M., Dhillon, A. P., Thomson, M. A., Harvey, P., Valentine, A., Davies, S. E., & Walker-Smith, J. A. (1998). Ileal-lymphoid-nodular hyperplasia, non-specific colitis, and pervasive developmental disorder in children. *Lancet (London, England), 351*(9103), 637–641. https://doi.org/10.1016/s0140-6736(97)11096-0 (Retraction published Lancet. 2010 Feb 6;375(9713):445)

Watkins, E. E., Zimmermann, Z. J., & Poling, A. (2014). The gender of participants in published research involving people with autism spectrum disorders. *Research in Autism Spectrum Disorders, 8*(2), 143–146. https://doi.org/10.1016/j.rasd.2013.10.010

Williams, I., Ayorinde, A. A., Mannion, R., Skrybant, M., Song, F., Lilford, R. J., & Chen, Y. F. (2020). Stakeholder views on publication bias in health services research. *Journal of Health Services Research & Policy, 25*(3), 162–171. https://doi.org/10.1177/1355819620902185

Appendix 2.1

Sex and Gender Equity in Research (SAGER) guidelines

Section of the article	Recommendations per section of the article
General principles	The authors should use the terms sex and gender carefully in order to avoid confusing both terms.
	Where the subjects of research comprise organisms capable of differentiation by sex, the research should be designed and conducted in a way that can reveal sex-related differences in the results, even if these were not initially expected.
	Where subjects can also be differentiated by gender (shaped by social and cultural circumstances), the research should be conducted similarly at this additional level of distinction.
Title and abstract	If only one sex is included in the study, or if the results of the study are to be applied to only one sex or gender, the title and the abstract should specify the sex of animals or any cells, tissues and other material derived from these and the sex and gender of human participants.
Introduction	The authors should report, where relevant, whether sex and/or gender differences may be expected.
Methods	The authors should report how sex and gender were taken into account in the design of the study, whether they ensured adequate representation of males and females, and justify the reasons for any exclusion of males or females.
Results	Where appropriate, data should be routinely presented disaggregated by sex and gender. Sex- and gender-based analyses should be reported regardless of positive or negative outcome. In clinical trials, data on withdrawals and dropouts should also be reported disaggregated by sex.
Discussion	The potential implications of sex and gender on the study results and analyses should be discussed. If a sex and gender analysis was not conducted, the rationale should be given. The authors should further discuss the implications of the lack of such analysis on the interpretation of the results.

Source: Heidari et al. (2016).

Appendix 2.2

Recommendations for sex and gender reporting in health research

Two paradigmatic shifts needed to fundamentally improve the quality of sex/gender and health research:

1 Sex/gender should not only be recognised, but also understood as intersecting with other axes of inequality such as race, ability, socioeconomic status, geographic location, sexual orientation, and age.
2 Gender should be conceptualised as a structural/social determinant of death, and should accompany any investigation of "sex" differences – in other words, research should not assume or proceed with the idea that "sex" can be separated from gender.

Six questions to help operationalise these paradigmatic shifts:

1 Does the study automatically give primacy to sex/gender? Does it move beyond asking whether sex/gender considerations are taken into account to explaining what relevant factors are taken into account to understand a particular illness, disease or health experience?
2 How does the study (biomedical, clinical, health systems, or population health focused) identify relevant factors that shape and determine health (e.g., ethnicity/race, sex/gender, age, socioeconomic status, geographic location, and sexual orientation)? What are the inclusion/exclusion criteria in relation to this question?
3 How does the research design (data collection and analysis) capture the relationships and interactions (e.g., using a multi-level analysis linking individual experiences to broader social structures) among pertinent health determinants and factors, including, but not limited to, sex/gender? Is the sample size adequate for capturing diversity between and within groups often treated in homogeneous manner (e.g., women and men)?
4 Does the study conceptualise and/or model gender as a social/structural determinant of health?

   a If yes, how?
   b If no, has a strong rationale been provided for how/why a gender conceptualisation is not needed – even if the researcher was not able to directly test the gender mechanism?

5 Does the study assert male/female differences in health related to biological mechanisms?

   a If yes, how are those biological mechanisms specifically explained and/or tested? Also, has it been explicitly described how gender and other intersecting factors are intertwined with these biological mechanisms?

b If no, does the study specifically state/demonstrate that intersecting social processes can cause the same biological mechanisms leading to male/female differences in health?

6 Where relevant, does the study contextualise research findings undertaken with human subjects within broader social structures and processes of power?

Source: Hankivsky et al. (2018).

Appendix 2.3

Checklist for gender-sensitive reporting

Component	Question	Responses	
		Yes	No
Research approaches	Are the concepts of gender and/or sex used in your research project?	□	□
	If yes, have you explicitly defined the concepts of gender and/or sex? Is it clear what aspects of gender and/or sex are being examined in your study?	□	□
	If no, do you consider this to be a significant limitation? Given existing knowledge in the relevant literature, are there plausible gender and/or sex factors that should have been considered? If you consider sex and/or gender to be highly relevant to your proposed research, the research design should reflect this.	□	□
Research questions and hypotheses	Does your research question(s) or hypothesis/es make reference to gender and/or sex, or relevant groups or phenomena (e.g., differences between males and females, differences among women, and seeking to understand a gendered phenomenon such as masculinity)?	□	□
Literature review	Does your literature review cite prior studies that support the existence (or lack) of significant differences between women and men, boys and girls, or males and females?	□	□
	Does your literature review point to the extent to which past research has taken gender or sex into account?	□	□
Research methods	Is your sample appropriate to capture gender and/or sex-based factors?	□	□
	Is it possible to collect data that are disaggregated by sex and/or gender?	□	□
	Are the inclusion and exclusion criteria well justified with respect to sex and/or gender? (Note: this pertains to human and animal subjects and biological systems that are not whole organisms).	□	□
	Is the data collection method proposed in your study appropriate for investigation of sex and/or gender?	□	□
	Is your analytic approach appropriate and rigorous enough to capture gender and/or sex-based factors?	□	□
Ethics	Does your study design account for the relevant ethical issues that might have particular significance with respect to gender and/or sex? (e.g., inclusion of pregnant women in clinical trials).	□	□

Source: Heidari et al. (2016).

Appendix 2.4

Seven strategies to improve the replicability of a systematic review using search engines

1 Researchers conducting systematic reviews should be explicit about the methodology they use to ensure sufficient consistency and repeatability. A detailed description should include the search platform used, the exact database used if search platform covers multiple databases, search date and time, the exact search strings, as well as whether the same search was replicated by more than one person. The locality/institution network from which the search was conducted should also be reported, preferably along with the IP address of the computer the queries were initiated from. Since even Web of Science's Core Collection consists of several sister databases, the precise reporting of the queried database should become common practice. The exact time of the search or the time window of the query are also essential. The holdings of databases, however, are not constant, historical records can be added over time, and, therefore, queries even within a clearly limited time period can deliver different result sets. Thus, reporting the time window of the queries can provide only a partial solution.

2 The use of adequate search platforms for a particular task should be an important consideration. All of the large platforms have different strengths; Google Scholar searches grey literature, Web of Science has the largest (combined) dataset, and, as our study confirmed, that Scopus and PubMed are the most consistent. Moreover, some databases may be more suitable for collecting information on a particular topic or have a greater historical coverage than others. In some countries, local search engines/databases may perform well for multiple criteria.

3 Peer reviewers and journal editors have an important role in safeguarding the repeatability reviews by enforcing precise reporting according to already established criteria.

4 Providers of scientific search platforms should consider opening their search code and relaxing their paywalls to make the full list of references resulted from a search publicly available, thus contributing to search transparency and, hence, scientific repeatability. Particularly Web of Science, as probably the most commonly used search platform, should act on making its search hits equally reachable to all users and, rather than a priori filtering them according to the institutions' paywall, restrict access only *after* the primary result set has been provided to the user.

5 Since Google Scholar has been criticised by the scientific community for the obscurity of its search algorithms, it could increase transparency in this regard to allow researchers to understand how the hit results are generated and how these are ordered. We acknowledge the business imperative but the need for research rigour is an important public good and facilitating this would enhance social licence.

6 Providing well-documented, standard application programming interfaces (APIs) would be greatly beneficial for researchers. These APIs could generate unique identifiers for searches and combine search term, result list, search time and location, and additional metadata (e.g.,

computing environment). Using an API for standardised searches would be particularly beneficial for searches using Google Scholar that shows a strong dependence on the computing environment. Although this solution could control for a great deal of variation derived mostly from computing background and would be able to keep detailed records on the metadata of the searches, it also brings up novel challenges. Firstly, APIs are admittedly more complex in terms of functionality and also in their use (which often needs some programming knowledge) than simple web interfaces. These may discourage users. Moreover, collecting detailed data about search locations, or even computing environment, raises both security and privacy concerns. Finally, storing individual searches along with the necessary metadata may be resource heavy, which is likely to increase maintenance costs, and therefore the subscription fees, of these services.

7 Alternatively, systematic review authors could deposit full list of their retrieved papers in open repositories as it is often done with raw data in many research areas. Alongside of these search outputs, metadata in a standard (machine readable) format about the search environment could be saved and deposited in these repositories. Web of Science, for instance, allows users to save search histories in *.wos files which, beside the search term, contain the exact queried databases. More studies are needed to confirm if using restricted databases provides a higher consistency in hit results among institutions.

Source: Pozsgai et al. (2021).
Note: References have been removed.

3 Journalology

The science of publishing

Abstract

Journalology is the study of any aspect of the academic publishing process, including the peer review process and the impact of predatory publishers on research dissemination. Some of the main aspects of journalology are presented in this chapter. It begins with a description of the historical origins of journalology, followed by an explanation of the peer review process. Common problems with the peer review process and potential solutions are then outlined. The emergence and impacts of preprinted articles and predatory publishers on the quality and dissemination of research are then explored. An explanation about how scholars can identify and mitigate the impact of predatory publishers concludes this chapter.

Keywords: Peer Review; Predatory Publishers; Preprinted Articles

Key points

- There are different ways that a manuscript can be peer reviewed. For example, single-blind peer review occurs when the peer reviewer knows the identity of the author, but the author does not know who is reviewing their paper. In contrast, double-blind peer review occurs when the author and the peer reviewer do not know the others identify. the others identify.
- The peer review process contains many flaws that can undermine the development and certification of credible research. For example, most peer reviewers are incapable of examining a manuscript that contains multidisciplinary concepts. To overcome this problem, manuscripts with multidisciplinary concepts can be sliced and peer reviewers can examine the concepts applicable to their expertise instead of evaluating the entire manuscript and commenting on unfamiliar concepts.
- Typically, a peer reviewer's knowledge of a discipline is deemed by the journal to be sufficient for them to evaluate a manuscript. However, such an assumption is not a substitute for experience. To support inexperienced peer reviewers, research teams should give them the opportunity to peer review a manuscript.
- To streamline and quicken the peer review process editors and peer reviewers should use the *Enhancing the QUAlity and Transparency Of health Research* (EQUATOR) and JBI checklists to ensure that the manuscript contains all the essential elements. Additionally, if a manuscript is resubmitted to another journal, after being rejected by a previous journal,

DOI: 10.4324/9781003510376-3

the author should also submit all previous peer review reports and their responses so that the editor can make a quick and accurate decision about if it should be peer reviewed for their journal.

- Preprinted manuscripts are published before they undergo peer review. Some researchers are unable to distinguish preprinted manuscripts from manuscripts that have been peer reviewed. Consequently, some credible scholarship has been corrupted by unexamined preprinted manuscripts. To avoid this problem, preprinted manuscripts should be clearly marked.
- Predatory publishers charge publication fees and do not subject the manuscript to a comprehensive peer review. To prevent the dissemination and proliferation of inferior manuscripts scholars should not cite in their research any manuscripts that have been published in predatory journals.

3.1   A description and historical origins of journalology

The term 'journalology', otherwise known as 'publication science', was first coined in 1989 by Stephen Lock, the former editor-in-chief of *BMJ* (i.e., formerly the *British Medical Journal*). Journalology is concerned with examining any aspect of the academic publishing process. It has been defined as 'the science of publication practices and the study of these activities' (Krishan & Kanchan, 2019, p. 1259). People working in journalology evaluate the peer review process and suggest reforms. They also study the impact of preprinted articles on the dissemination of research and explore the consequences of predatory publishers. These aspects of journalology are outlined in this chapter.

3.2   The peer review process

The peer review process began during the early to mid-17th century within higher learning establishments in Western Europe. This process remained virtually unchanged until the publication of manuscripts became an essential requirement for obtaining tenure or promotion in American higher education institutions during the twenty-first century (Hoffman, 2022). It is most likely that these requirements led to the peer review process becoming a billion-dollar enterprise (LeBlanc et al., 2023). Kovanis and colleagues (2016) estimated that in 2015, 1.1 million journal articles were published on Medline, which required about 9 million reviews and about 1.8 million peer reviewers. In another study, Aczel and colleagues (2021) reported that in 2020, academics devoted more than 100 million hours globally to examining academic articles (see Figure 3.1). They also reported that in the United States of America in 2020 USD $1.5 billion worth of labour was used to peer review manuscripts. In comparison, peer reviewing activities in China was USD $600 million and in the United Kingdom this amount was USD $400 million. The figure in illustration 3.7 calculates the labour hours using a nominal six hours per article per reviewer, although it is likely that shorter articles may take less time and lengthy articles may take more time. The figure assumes two peer reviewers for rejected articles and three for acceptance, though some articles will only have one or two reviewers, whether accepted or rejected.

The peer review process involves editors, peer reviewers, and manuscript authors. This process begins when a manuscript is submitted to an academic journal. The journal's editor then examines its contents and decides if it should be rejected because it is poorly written and/

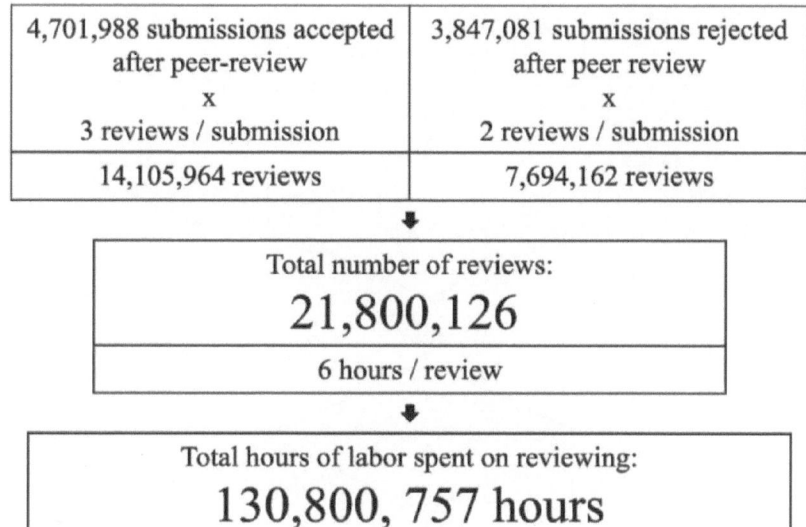

4,701,988 submissions accepted after peer-review	3,847,081 submissions rejected after peer review
x	x
3 reviews / submission	2 reviews / submission
14,105,964 reviews	7,694,162 reviews

Total number of reviews:
21,800,126
6 hours / review

Total hours of labor spent on reviewing:
130,800, 757 hours

Figure 3.1 Overview of time spent peer reviewing scholarly articles in 2020
Source: Aczel et al. (2021, p. 4).

or will not be suitable for the journal's readership. Alternatively, if the editor deems it to be potentially suitable then they invite potential peer reviewers to examine its contents. There are four types of peer review. First, single-blind peer review occurs when the peer reviewers know the identity of the author, but the author is unaware of who is reviewing their manuscript. Second, double-blind peer review occurs when the author and the peer reviewer do not know the other's identity. Third, open peer review occurs when the authors and peer reviewers know each other's identities. Open peer review also occurs when the manuscript is accepted for publication, and the peer reviewer's comments and the author's subsequent responses are published along with the manuscript. Fourth, transparent peer review occurs when the peer reviewers know the identity of the author, but the author does not know the identities of the peer reviewers unless the peer reviewers choose to announce their identity. Once a manuscript has undergone the peer review process the journal's editor then evaluates the comments and either decides to publish the manuscript unaltered or invites the author to make changes as recommended by the peer reviewers. Once the author makes the appropriate alterations or successfully refutes the peer reviewers' suggestions the manuscript is published (see Figure 3.2).

Glonti and colleagues (2019) identified 76 role-related statements and 73 task-related statements that peer reviewers were expected to perform when examining manuscripts submitted to biomedical journals. For a successful peer review to occur peer reviewers must be proficient in the essential tasks required to evaluate and decide if a manuscript should be published. As illustrated in Figure 3.3, the main tasks that peer reviewers performed were to assess the contents of each section of the manuscript (i.e., methods, results, discussion, conclusion, and references). Peer reviewers were also expected to assess the manuscript's presentation and scrutinise ethical aspects of the study (see Figure 3.3) (Glonti et al., 2019).

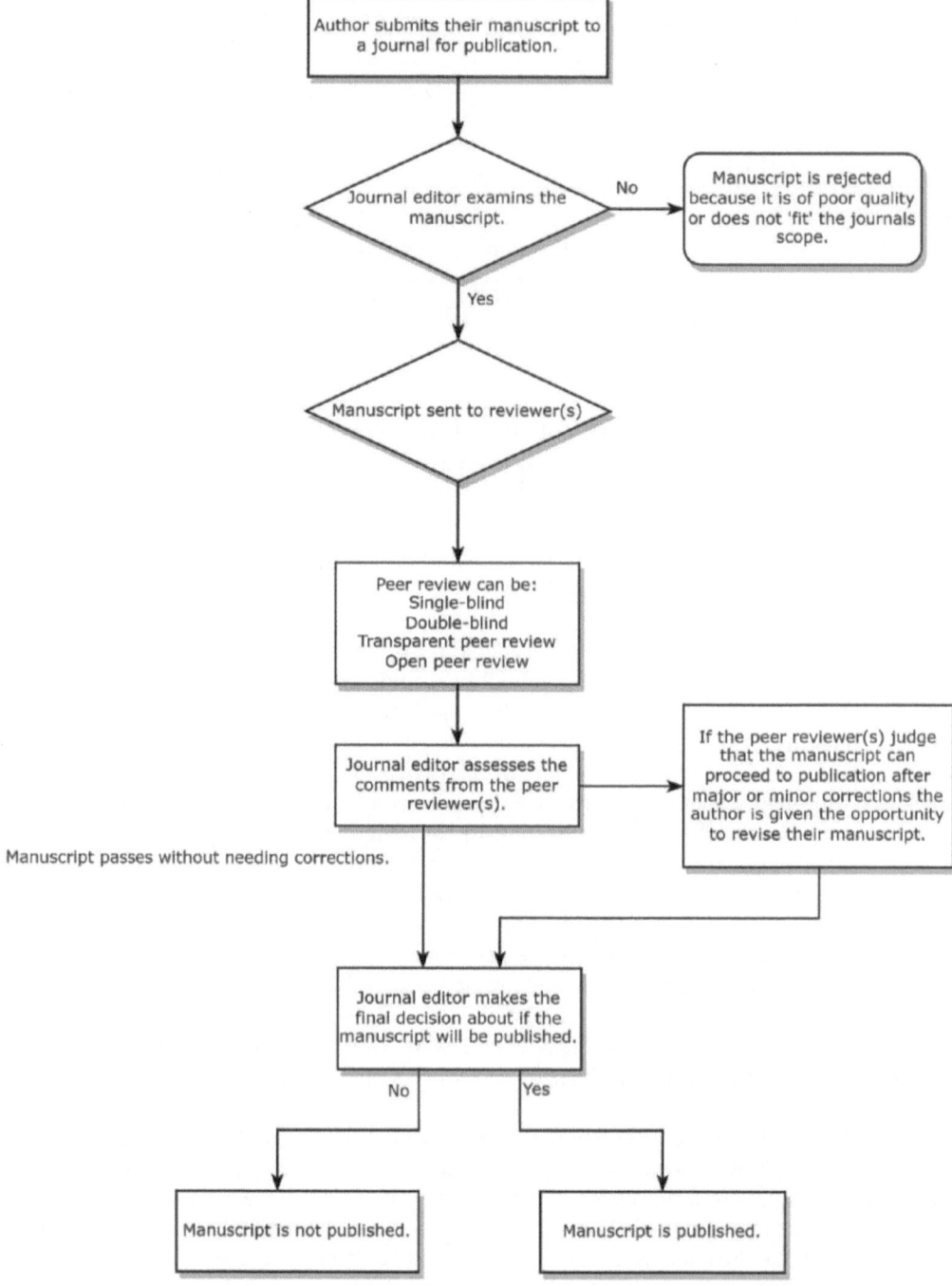

Figure 3.2 The peer review process

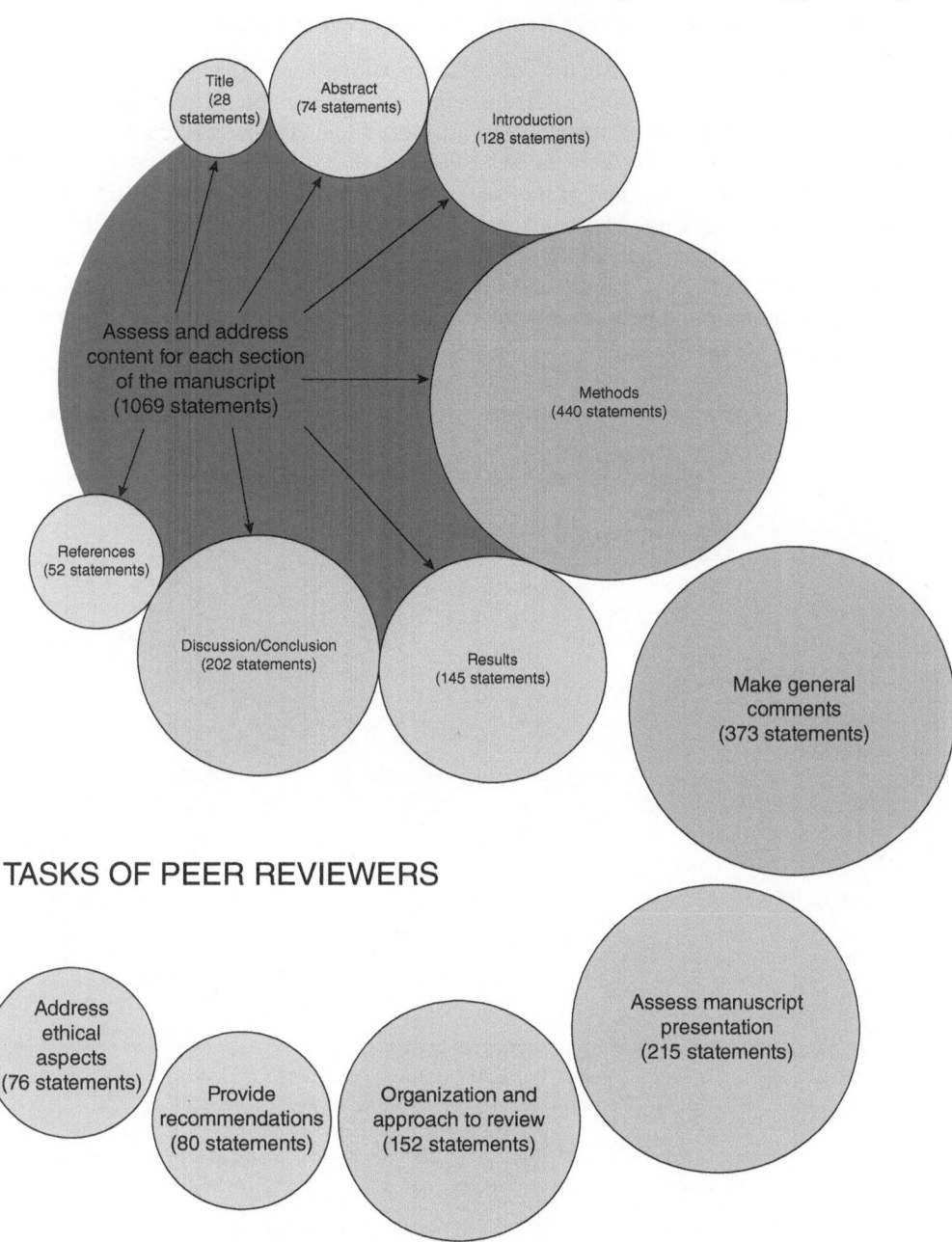

Figure 3.3 Themes related to tasks of peer reviewers

Source: Glonti et al. (2019, p. 12).

3.3   Issues and potential improvements of the peer review process

For the past several centuries, peer review has been the cornerstone of evaluating the accuracy of academic research. However, it is not a flawless process. For example, publication bias occurs when manuscripts that contain appealing content are more inclined to be selected for peer review and possibly published. Appealing content may be content that is novel or focused on topics that are currently a focus for funders and/or the media, such as research around COVID-19 during the COVID-19 pandemic. Heim and colleagues (2018) have illustrated four main difficulties with the peer review process (i.e., which are described in the main square shapes) along with some relevant solutions (i.e., which are described in the circles) (see Figure 3.4). Some of the problems and solutions to the peer review process that Heim and colleagues have outlined are elaborated in this section.

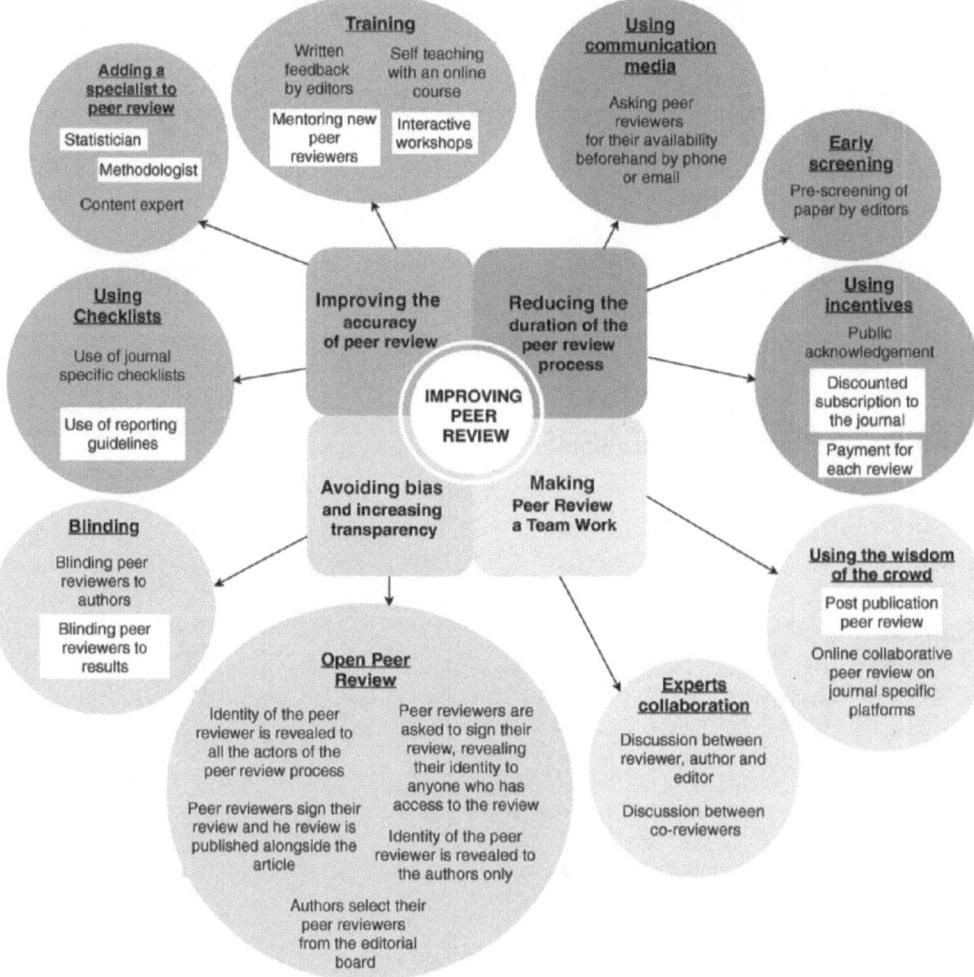

Figure 3.4 Problems and solutions to the peer review process as outlined by Heim and colleagues (2018)

Source: Heim et al. (2018, p. 3).

3.3.1   *Making peer review a teamwork effort*

3.3.1.1   *Segmented peer review*

Science's gatekeepers are editors and peer-reviewers with the disciplinary, methodological, and content expertise to ensure quality control. However, the current peer-review system was designed centuries ago when science consisted of single authored projects. Today's more complex, translational health science environment calls for a review process with a more expansive and diversified expertise—one that is commensurate with the diversified skills and knowledge of multidisciplinary teams of authors. No matter how well-intentioned, two peer-reviewers are unlikely to have that capacity.

(Smith, 2021, p. 1219)

Dinakaran and colleagues (2021) have proposed that manuscripts that contain multidisciplinary concepts could be examined using a segmented peer review process. This process begins when the author informs the journal that their manuscript contains multidisciplinary concepts. The editor then recruits suitable experts who will only examine the section of the manuscript that aligns with their knowledge and experiences. Once each reviewer has examined their allocated section, they then inform the editor as to the academic merit of the examined content. The editor then integrates each reviewer's recommendations and decides if the manuscript should be rejected or published with or without revisions (see Figure 3.5).

When confronted with a manuscript that contains multidisciplinary concepts, the editor might ask the manuscript's author to nominate several possible peer reviewers. Despite this practical approach, some literature has concluded that author-suggested peer reviewers provide more lenient assessments of the manuscript compared to peer reviews approached by editors (Bornmann & Daniel, 2010; Kowalczuk et al., 2015; Liang, 2018; Shopovski et al., 2020). To avoid this potential problem, editors should only accept author-suggested peer reviewers

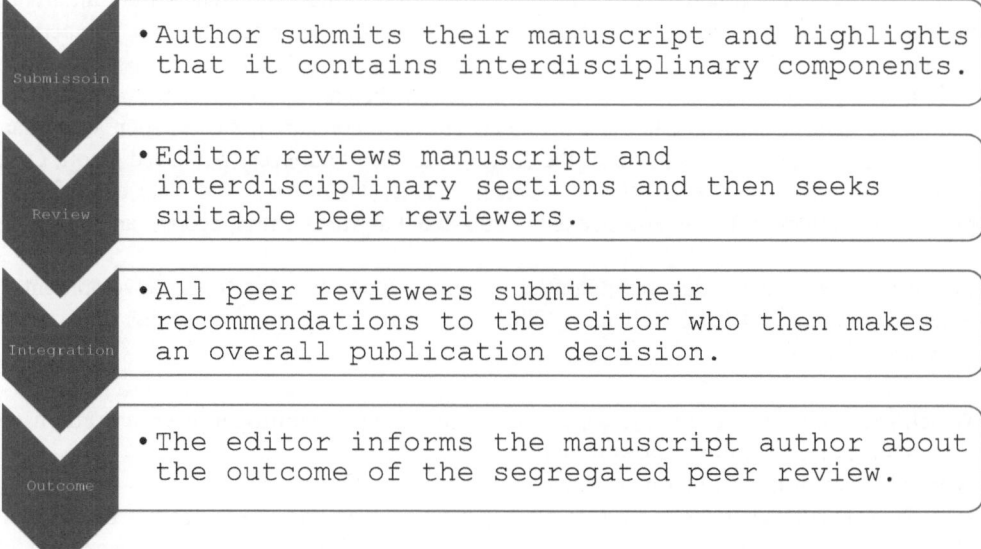

Figure 3.5 Components of a segmented peer review

provided the author has not informed the potential peer reviewer that they have been nominated. This confidentiality can ensure that the peer reviewer is unaware of the manuscript's author and will therefore not provide a more favourable or biased assessment.

3.3.1.2   *Expert collaboration*

One condition of peer reviewing a manuscript is that the peer reviewers are not permitted to share it with other experts, even though sharing it could improve their assessment. One rationale for not sharing is so the ideas examined cannot be used by other scholars. Due to this confidentiality only the peer reviewer's judgement, which can be biased and/or limited, is used to evaluate the manuscript. To improve the peer review process once a journal publishes a manuscript they can use a post-publication peer review process, whereby experts examine the manuscript and debate whether it should remain in circulation (Heim et al., 2018). Such an approach has the potential to reduce the time it takes to retract flawed studies and the prospect of flawed research being cited.

3.3.2   *Avoiding publication bias and increasing transparency*

3.3.2.1   *Open peer review*

The conventional peer review process is defined by three features. First, this process is either partially or completely anonymous (Schmidt et al., 2018). For example, a double-blind peer review occurs when neither the peer reviewers nor the manuscript's authors are aware of each other's identities (Tomkins et al., 2017). Second, peer reviewers are selected by editors (Ross-Hellauer et al., 2017). Third, recommendations from peer reviewers and the editor's decision are not publicised (Schmidt et al., 2018). In the interests of avoiding any real or perceived conflicts of interest journals can implement an open peer review system. Heim and colleagues (2018, p. 4) have described this system:

> Manuscripts are posted online on an open access platform where researchers from all around the world with any background can peer review the study. Chosen researchers are also actively invited by the author and the editor to peer review the online publication. The peer review is entirely transparent: the reviewers' names and affiliation, their report and the approval status they choose are published along with the article. Peer review reports are posted as soon as they are received and the peer review status of the article is updated with every published report. Once an article has passed peer review (i.e., it has received at least two 'Approved' statuses from independent peer reviewers), it will be indexed in PubMed, PubMed Central, Scopus, and Embase.

Although open peer review has the potential to deliver many benefits to the development of knowledge, Tennant and Ross-Hellauer (2020, p. 9) have posed some questions that remain unanswered, such as:

1   Which specific open peer review systems (run via journals or third-party services) do users (within differing disciplines) most prefer?
2   What measures might further incentivise uptake of open peer review?
3   How fixed are attitudes to the various facets of open peer review and how might they be changed?
4   How might shifting attitudes towards open peer review impact willingness to engage with the process?
5   How have attitudes changed over time? As open peer review gains familiarity amongst researchers and is further adopted in scholarly publishing, do attitudes towards specific elements like open identities change? In what ways?

6 To what extent are attitudes and practices regarding open peer review consistent? What factors influence any discrepancies?

7 Is an openly participatory process more attractive to reviewers, and is it more effective than traditional peer review? And if so, how many participants does it take to be as or more effective?

8 Does openness change the demographic participation in peer review, for authors, editors, and reviewers?

3.3.2.2   Transparent peer review

One criticism of open peer review is that academics are inclined to reject the invitation to review a manuscript because their reviews and identities will be published. One possible solution to this dilemma is 'transparent peer review', whereby the peer reviewer's report is published anonymously. Currently, *Nature Communications* and *Genome Biology* use transparent peer review (Cosgrove & Cheifet, 2018; Nature Communications, 2016). Cosgrove and Cheifet (2018) compared the publishing characteristics of 110 submissions to *Genome Biology*; 45 submissions that underwent transparent peer review and 65 submissions that underwent single-blind peer review (i.e., the peer reviewers know the authors' identities, the peer reviewer's identity remains anonymous to the authors, and the peer reviewer's assessment of the manuscript is not published). They reported that between transparent and single-blind peer review processes, there were no significant differences in the average time from submission to a publication decision or the number of peer reviewers invited to examine the manuscript. Based on these findings, Cosgrove and Cheifet (2018, pp. 1–2) concluded that:

> One criticism of traditional peer review is that a reader does not know who the reviewers were, and so cannot judge whether they had sufficient expertise to assess the manuscript, thus meaning the stamp of 'peer reviewed' is of uncertain value. Although transparent review does not fully address this, since the reader still will not know the reviewers' identities, the reader will be able to read the comments and assess whether the reviewer has made sensible and reasonable criticism of the work, which should lead to increased confidence in the peer-review assessment. Fully open review would be even better from this point of view, but many reviewers may be reluctant to associate their names with negative reviews, even when the negative comments are justified, for fear of retaliation. It seems to us that transparent review is currently the best compromise, and we hope that, when it is well established, it will foster a more open environment where reviewers will feel more comfortable in revealing their identities.

3.3.2.3   Withholding results from the peer review process

As mentioned earlier, publication bias occurs when a manuscript is selected for peer review because it has appealing features, such as novel and/or exciting results. Due to this bias more methodologically rigorous research that is not as appealing is often not subjected to peer review. This view was once expressed by Chambers (2014), who stated:

> publication bias is simple human nature: in judging whether a manuscript is worthy of publication, editors and reviewers are guided not only by the robustness of the method but by their impressions of what the results contribute to knowledge. Do the outcomes constitute a major advance, worthy of space within a journal that rejects the majority of submissions? Results that are novel and eye-catching are naturally seen as more attractive

and competitive than those that are null or ambiguous, even when the methodologies that produce them are the same.

(Kretser et al., 2019, p. 346)

As illustrated by Ayorinde and colleagues (2020), publication bias occurs between the manuscript being submitted to the journal for consideration and the peer review and editorial decision about the manuscript (see Figure 3.6).

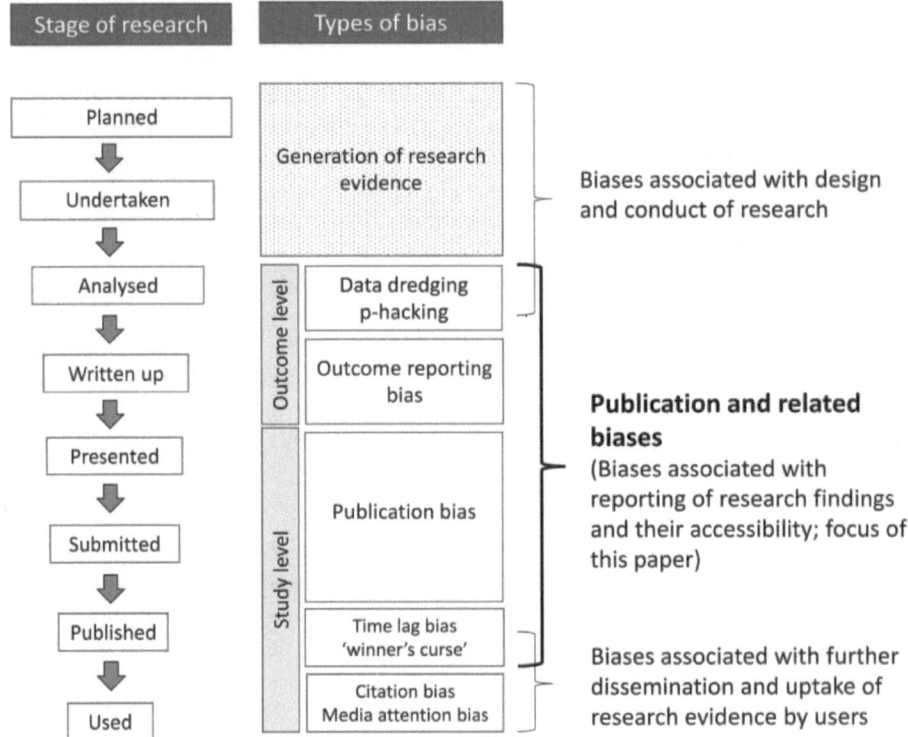

Figure 3.6 Biases at different stages of the research process

Source: Ayorinde et al. (2020, p. 2).

A 'results-free' manuscript submission process has been developed to reduce the occurrence of publication bias. With this submission process, a detailed description of the study's design is submitted for peer review. Peer reviewers then determine if it is methodologically appropriate. Peer reviewers can either reject the study's design, suggest modifications, or recommend that the study should proceed unamended. If the study's design is appropriate then the author conducts the study and submits a complete study for final peer review, which has both the approved study design and its results (Button et al., 2016). The 'results-free' manuscript submission process has been embraced by *BMC Psychology* because:

> The results-free review model, launched this month in BMC Psychology, offers a solution by focusing editorial decisions on the scientific rigour of the study design, and preventing editorial decisions being unduly biased by study findings. The human powers of self-persuasion and post-hoc justification mean that withholding results from peer-reviewers may be the only reliable way to protect reviewers and editors against the often unconscious influence of the results justifying the means.

(Button et al., 2016, p. 6).

3.3.3   *Improving the accuracy of the peer review process*

3.3.3.1   *Training and opportunities to peer review manuscripts*

Typically, the only reason why peer reviewers are invited to review a manuscript is because they are deemed to possess expert knowledge about the concepts explained in the manuscript. Consequently, their proficiency in examining other aspects of the manuscript is often overlooked. This sentiment was once expressed by Glonti and colleagues (2019, p. 2), who stated:

> Unlike other professional groups, many editors and peer reviewers of biomedical journals operate largely without formal training. It is assumed that having expertise as an author provides, by default, the skills necessary to be a scientific editor and/or peer reviewer. However, this assumption is problematic, potentially having a number of negative implications for the overall quality of biomedical publishing.

Training courses can give peer reviewers the opportunity to refine their ability to conduct a complete and competent examination of the manuscript. Galica and colleagues (2018) created a two-day curriculum that can be used to teach biomedical students the skills required to perform a satisfactory peer review. Along with providing training, prospective peer reviewers should be given multiple opportunities to conduct peer reviews (Dennehy et al., 2021). Such chances can give them the experience they need to refine their peer review skills and a sense of being a valued member of the academic community. This sentiment was articulated by John Dennehy, who stated:

> I was never asked to participate in a peer review by my mentors. Being asked to review a manuscript was an important milestone for me as it indicated that I was accepted as a peer in the scientific community. However, it took some time for me to learn the ins and outs of reviewing, a process that would have been facilitated if my mentors had shared reviewing responsibilities with me. The peer review process is a fundamental aspect of science. Despite this, specific training in the performance of peer review is rare. There is more to peer review than simply critiquing a paper. Not only must the scientific soundness of the work need to be judged, but also the significance of the work, its context in the literature, and its appropriateness for the journal need to be considered. As a mentor, I include my mentees in all reviews that I perform. The mentee benefits from first-hand exposure to the peer review process and the authors and editors benefit from an extra set of eyes assessing the work. But too often the contributions of "junior" reviewers are not acknowledged. Direct recognition of their involvement will benefit trainees' careers by increasing their recognition as scientists by the scientific community.
>
> (Dennehy et al., 2021, p. 2)

3.3.4   *Reducing the duration of the peer review process*

3.3.4.1   *Early screening of manuscripts*

Peer reviewers sometimes waste their time, efforts, and resources evaluating inferior manuscripts. To avoid this situation, as a condition of publication, the author should submit the most appropriate *Enhancing the QUAlity and Transparency Of health Research* (EQUATOR) checklist along with their manuscript. These checklists can help ensure that only high-quality manuscripts are submitted for peer review. Struthers and colleagues (2021) have published a flow diagram that authors, peer reviewers, and editors can use when deciding the most appropriate EQUATOR checklist to use for evaluating a manuscript (see Figure 3.7). Alternatively, if the

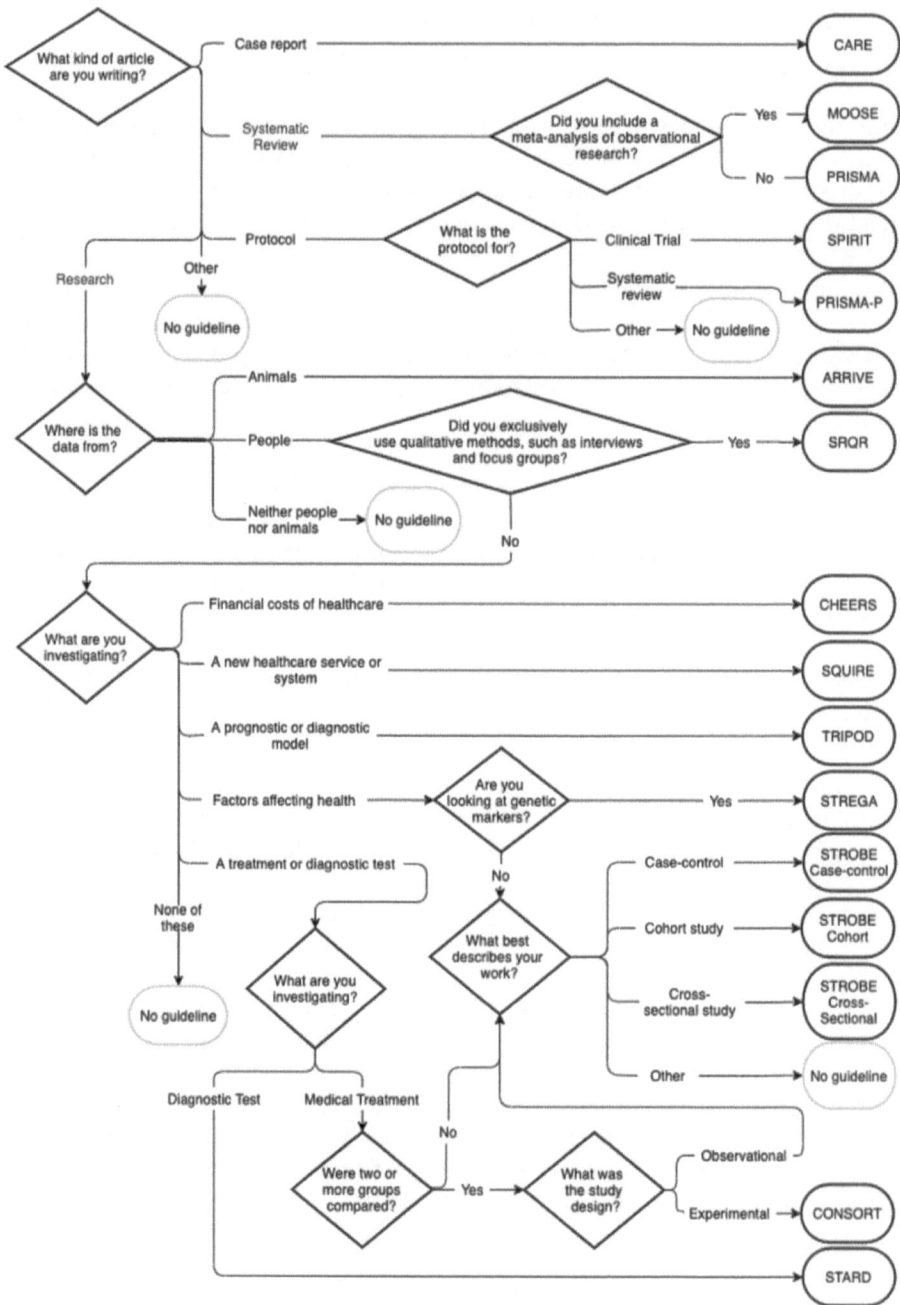

Figure 3.7 EQUATOR reporting guideline decision tree

Source: Struthers et al. (2021, p. 4).

Acronyms: ARRIVE, Animal Research Reporting of In Vivo Experiments; CARE, CAse Report; CHEERS, Consolidated Health Economic Evaluation Reporting Standards; CONSORT, CONsolidated Standards of Reporting Trials; MOOSE, Meta-analysis Of Observational Studies in Epidemiology; PRISMA, Preferred Reporting Items for Systematic Review and Meta-Analysis Protocols; SPIRIT, Standard Protocol Items: Recommendations for Interventional Trials; SQUIRE, Standards for QUality Improvement Reporting Excellence; SRQR, Standards for Reporting Qualitative Research; STARD, Standards for Reporting Diagnostic Accuracy; STROBE, STrengthening the Reporting of OBservational studies in Epidemiology; TRIPOD, Transparent Reporting of a multivariable prediction model for Individual Prognosis Or Diagnosis

EQUATOR checklists are unavailable or unsuitable for usage editors could mandate that authors submit with their manuscript a relevant JBI checklist (see Appendix 3.1).

3.3.4.2   *Using incentives to entice peer reviewers*

Currently, peer reviewers are rarely rewarded for evaluating manuscripts and when they examine a manuscript they rarely devote sufficient time or effort to this exercise due to competing academic commitments. Due to this lack of incentive when approached some are not interested in peer reviewing a manuscript. However, this trend is starting to change with some journals now offering a discount subscription to the journal or an honorarium payment for their assessment (Heim et al., 2018). Along with these incentives, Kretser and colleagues (2019, p. 344) have suggested that:

> serving as a reviewer should be a role that is built into career advancement. To a minor extent, this is already being implemented for tenure-earning faculty at some universities. This benefit provides an incentive for scientists to participate and further ensures that those who are experts in their field will be peer reviewers.

3.3.5   **Other suggestions to improve the peer review process**

3.3.5.1   *Removing multiple examinations in the peer review process*

There are many suggestions about how prospective authors can successfully navigate their manuscripts through the peer review process (Agathokleous, 2022; Annesley, 2011; Baker et al., 2017). One common recommendation proposed is that authors should incorporate the peer reviewer's suggestions only if they do not jeopardise the intellectual integrity of the arguments they are explaining in their manuscript (Agathokleous, 2022). However, evidence suggests that authors are more inclined to make some minor adjustments to their manuscript and then submit it to another journal (Crijns et al., 2021). Crijns and colleagues (2021) reported that of the 250 rejected manuscripts that they examined, 200 (80%) were published in another journal. Among the 609 substantive, actionable items identified in the rejection letters of the 200 manuscripts that were eventually published, 205 (34%) were addressed in the published manuscripts. Based on these results, Crijns and colleagues (2021, p. 517) concluded that:

> Our findings suggest that authors often disregard advice from peer reviewers after rejection. Authors may regard the peer review process as particular to a journal rather than a process to optimize dissemination of useful, accurate knowledge in any media.

Crijns and colleagues have shown that authors tend to disregard suggestions proposed by peer reviewers and just submit the same manuscript to another journal. According to Bennett and Goodall:

> The act of submitting a manuscript to another journal without first incorporating previous peer review suggestions can undermine the production of high-quality research. This act wastes the peer reviewer's time because they are having to examine a manuscript that has already been reviewed. It also ignores previous peer review suggestions that were valid.
>
> (Bennett & Goodall, 2022, pp. 192–193)

To prevent this behaviour from occurring Crijns and colleagues suggest that journals should use a single manuscript submission website that facilitates the transferring of peer review reports from one journal to another. Alternatively, Bennett and Goodall proposed that:

> To save a peer reviewer's time and to improve the quality of published manuscripts, a condition of publication should be that authors declare if their manuscript has been previously peer-reviewed. If so, they should also be obligated to provide all reports by the previous peer reviewers. This documentation can accelerate the peer review process because an Editor can make a quick and accurate decision about if it should be sent out to peer-review. Such documentation can also augment another peer reviewer's comments, thus resulting in a more comprehensive assessment of the manuscript and a better-quality study.
>
> (Bennett & Goodall, 2022, p. 193)

3.3.5.2   *Publishing manuscripts about the peer review process*

Historically, improvements to the peer review process often remained unknown to academics because such recommendations were dispersed throughout different disciplines. However, the academic journal *Research Integrity and Peer Review* rectified this situation because from across different academic disciplines recommendations about the peer review process have been amalgamated (Harriman et al., 2016). *Research Integrity and Peer Review* have published an eclectic range of articles about the peer review process (see Table 3.1) (Boughton et al., 2018). In the interests of improving the dissemination of knowledge that can ultimately enhance the peer review process other academic journals about this process should be created and their articles should be open access, so that they are available to all.

3.3.5.3   *Improving the professionalism of reports by peer reviewers*

Unprofessional comments by peer reviewers can increase the psychological distress of early career researchers. Gerwing and colleagues (2020) evaluated 1,491 sets of comments by peer reviewers who evaluated manuscripts published in the fields of behavioural medicine, ecology, and evolution. They reported that 179 sets of comments (12%) included at least one unprofessional comment by the peer reviewer towards either the manuscript's author or their work and 611 sets of comments (41%) had either incomplete, inaccurate, and unsubstantiated critiques. They concluded that a peer reviewer's unprofessional comments can exacerbate the psychological distress academics experience, especially those who are new to academia. In another study, Silbiger and Stubler (2019) distributed an anonymous survey to international students in science, technology, engineering, and mathematics (STEM) fields to measure the impacts of unprofessional comments from peer reviewers on authors. They concluded that marginalised groups in STEM fields were more inclined to report a reduction in their scientific aptitude, productivity, and career advancement after receiving a peer review report that contained unprofessional comments.

To support underrepresented groups in STEM and early career researchers, editors should ban peer reviewers who write unprofessional comments in their peer review reports. To help them decide if a comment was unprofessional, editors can use Silbiger and Stubler's definition of unprofessional comments. According to these authors an unprofessional comment:

1 Lacks constructive criticism about the manuscript,
2 Is directed to the author(s) rather than to the nature or quality of their work,
3 Use personal opinions about the author(s) manuscript instead of evidence-based criticisms, and
4 Are cruel and/or mean-spirited.

Table 3.1 Topics published within the journal research integrity and peer review

Journal section	Subject area (reference in brackets)
Research and publication ethics	Conflicts of interest disclosure (Dunn et al., 2016) Guidelines on research integrity (Räsänen & Moore, 2016) Costs of ethical review (Barnett et al., 2016) Citation bias (Duyx et al., 2017; van der Vet & Nijveen, 2016) Plagiarism (Higgins et al., 2016) Research misbehaviours (Bouter et al., 2016) Text recycling (Moskovitz, 2017; Roig, 2017) Reasons for retractions (Bozzo et al., 2017) Uses of expressions of concern (Vaught et al., 2017) Research ethics review (Page & Nyeboer, 2017) Research funding (Barnett et al., 2017) Author contributions (Boyer et al., 2017) Author perceptions of publishing (Wallach et al., 2018)
Research reporting	Reporting on sex and gender (Heidari et al., 2016; Welch et al., 2017) Standards of reporting (Hamilton et al., 2016; Korevaar et al., 2016; Shanahan et al., 2017) Data sharing (Hrynaszkiewicz et al., 2016; Rowhani-Farid et al., 2017) Trial registration (Asiimwe & Rumona, 2016; Gray et al., 2017) Readers perceptions on research (Wager et al., 2016) Factors associated with online media attention of research articles (Haneef et al., 2017)
Peer review	Training in peer review (Byrne, 2016; Scarrow et al., 2017) Reviewer recruitment (Albert et al., 2016; Fox et al., 2017) Mentoring in peer review (Wong et al., 2017) Views on peer review models (Patel et al., 2017) Peer review of grant proposals (Coveney et al., 2017)
General relevance	Proceedings of the 4th World Conference on Research Integrity (O'Brien et al., 2016) Proceedings from the IV Brazilian Meeting on Research Integrity, Science and Publication Ethics (Vasconcelos et al., 2017)

Source: Boughton et al. (2018, p. 2).

3.4   Preprinted articles

Generally, a manuscript undergoes a peer review process before it is published. In contrast, preprint manuscripts are published before they undergo a peer review process (see Figure 3.8) (Tennant et al., 2019).

There are five reasons why scholars publish preprint manuscripts. First, they may believe that because their manuscript contains important information, it should be published as a preprint before it undergoes peer review. Second, they would like their manuscript published, even if it does not pass the peer review process. Third, they would like to disseminate their manuscript to a broader audience along with publishing it in an open access journal. Fourth, their study may have negative results that they believe will not pass conventional peer review process. Fifth, they may be frustrated with the time required for peer review and may instead decide to distribute their study's results as a preprint manuscript (Elmore, 2018).

Some scholars are unable to distinguish a peer reviewed and preprint manuscript. Such confusion can result in scholars incorporating into their research unscrutinised findings that may later be proven to be false. To avoid such an outcome, the *American Medical Writers Association*, the *European Medical Writers Association*, and the *International Society for Medical Publication Professionals* published a joint policy statement about preprinted manuscripts

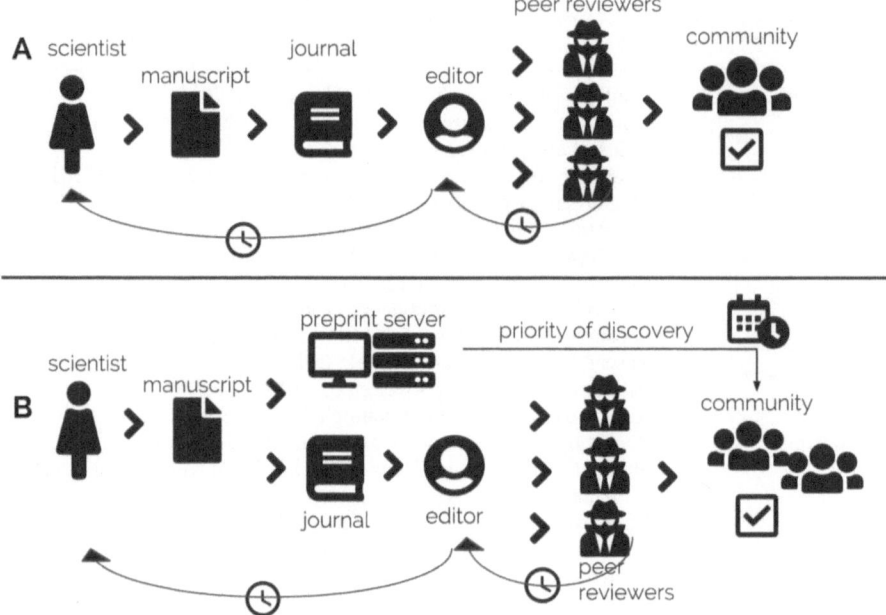

Figure 3.8 Publishing process for peer review (A) and preprint manuscripts (B)'
Source: Tennant et al. (2019, p. 3).

(American Medical Writers Association et al., 2021). Ravinetto and colleagues (2021) have also provided five recommendations to ensure that preprinted manuscripts can be distinguished from peer reviewed manuscripts (see Table 3.2).

Despite their popularity there are inconsistent policies about the usage of preprinted articles in academic journals. Klebel and colleagues (2020) evaluated the clarity of policies about peer review and preprinted articles in 171 major academic journals. They reported that 31.6% of journal policies did not provide any details about the type of peer review that the journal used and 39.2% of journal policies did not explain if preprint manuscripts could be published. Due to these findings, it is important that journals address this policy void and provide clarification as to if preprint manuscripts can be cited in manuscripts that are submitted for peer review.

3.5   The creation of predatory publishers and Beall's list

Prior to the widespread adoption of the internet, almost all academic journals were printed on paper and were only available to institutions with subscriptions to the publisher. Peer review during this time was strictly controlled, and the result was frequently the release of excellent research that could bear scrutiny. There were a few poor-quality academic publishers at this time. However, as most scholars were aware of their existence, they either refrained from submitting their manuscripts for publication or avoided citing the manuscripts of predatory publishers (Beall, 2017).

In North America, many college and university libraries started cancelling their journal subscriptions in the 1980s and 1990s since the cost of these subscriptions had increased but library funding had not. The price of journal subscriptions had increased for three reasons. First, many baby-boomers were approaching the stage of their lives when they were finishing up their PhDs and starting careers in academia where they were required to produce publications. As a result,

Table 3.2 Five recommendations by Ravinetto and colleagues (2021) about preprint manuscripts

Recommendation Number	Recommendation
1	Consensus should be sought on a term clearer than 'pre-print', such as "Unrefereed manuscript", "Manuscript awaiting peer review" or "Non-reviewed manuscript".
2	Caveats about unrefereed manuscripts should be prominent on their first page, and each page should include a red watermark stating 'Caution—Not Peer Reviewed'.
3	Pre-print authors should certify that their manuscript will be submitted to a peer-review journal, and should regularly update the manuscript status.
4	High level consultations should be convened, to formulate clear principles and policies for the publication and dissemination of non-peer reviewed research results.
5	In the longer term, an international initiative to certify servers that comply with good practices could be envisaged.

Source: Ravinetto et al. (2021).

journals' workload grew to keep up with the growth in research output, and some journals switched from biannual to quarterly publication. Second, the price of international journal subscriptions increased in the late 1990s due to the weaker Canadian and American currencies. Third, a variety of new disciplines appeared with the entry of the baby boomer generation into higher education, such as nanomaterials and genomics, which made it unprofitable for university libraries to pay subscription fees for specialised periodicals (Beall, 2017).

Most higher education institutions blamed the publisher's greed for price increases rather than rising costs related to the production of manuscripts. The rise of the open access movement coincided with this cost increase and the development of the internet. However, shortly after this trend started, pay-to-publish predatory publications started to arise. Beall became aware of them for the first time in 2008 when he began to get spam emails asking him to send them his work for consideration (Beall, 2017).

A strict peer review procedure is typically used by both subscription-based and open-access journals to guarantee that the publication is of high-quality. As a result, dubious articles are rarely published. Predatory publishers, however, are more concerned with charging an author an article processing fee instead of publishing high-quality research. They either have a poor or non-existent peer review process (Beall, 2016a). Beall (2017, p. 275), who articulated this viewpoint, stated that:

> What I learned from predatory publishers is that they consider money far more important than business ethics, research ethics, and publishing ethics and that these three pillars of scholarly publishing are easily sacrificed for profit. Soon after they first appeared, predatory publishers and journals became a godsend both for authors needing easy publishing outlets and sketchy entrepreneurs wanting to make easy money with little upfront investment.

In response to the establishment and proliferation of predatory publishers Beall (2016b) created four separate lists, which are typically referred to as 'Beall's list'. These four lists are defined below:

1 Predatory or questionable publishers;
2 Predatory or questionable journals;
3 Journals that imitate an already established reputable journals; and
4 Fake metrics companies (Beall, 2016b).

3.6   Consequences of predatory journals

3.6.1   Corrupting research

Predatory journals are more likely to disseminate flawed studies since the studies they publish have not received adequate peer review. Consequently, scholars might inadvertently cite in their own publications such flawed research. Consequently, the trustworthiness of their research, and the entire discipline in general, is corrupted. This sentiment has been articulated by Tsuyuki and colleagues (2017, p. 274), who claimed that:

> Peer review is the coin of the realm of science, and because predatory journals either carry out a fake peer review or are negligent at managing it, they often publish science that has not been properly vetted. And because research is cumulative, unscientific papers pollute the pool of published science (and evidence), threatening future research and making it difficult for clinicians to wade through the evidence.

To prevent the tarnishing of credible research, academics can use the strategies that Rice and colleagues (2021) have outlined to screen out predatory journals from their systematic literature reviews (see Table 3.3).

3.6.2   Undermining the training of scholars

> … since the advent of predatory publishing, there have been tens of thousands of researchers who have earned Masters and Ph.D. degrees, been awarded other credentials and certifications, received tenure and promotion, and gotten employment – that they otherwise would not have been able to achieve – all because of the easy article acceptance that the pay-to-publish journals offer.
>
> (Beall, 2017, p. 275)

As explained in the quotation above, some scholars have received their academic credentials and jobs after publishing research in predatory publishers. Academics who have built their careers on publishing flawed studies are likely to have inadequate research skills compared to scholars who have published their research with legitimate publishers. A flow-on effect is that they are unable to competently teach and mentor their successors (Beall, 2017). Unless this situation is corrected, it is plausible to argue that the upcoming generation of scholars will be less academically proficient than previous generations.

3.6.3   Increased email correspondence to academics

Some academics are inundated with spam emails from predatory publishers, which can distract them from focusing on their teaching and publishing activities (Krasowski et al., 2019; McKenzie et al., 2021; Sousa et al., 2021; Wood & Krasowski, 2020). Krasowski and colleagues (2019) examined the email inboxes and junk folders over seven consecutive days of 17 faculty staff (i.e., four assistants, four associates, and nine full professors) and nine trainees (i.e., five medical students, two pathology students, and two pathology fellows). In total, 755 emails met their eligibility criteria (i.e., 417 emails from 328 unique journals, 244 conference invitations, and 94 webinar invitations). They reported that full professors received the most emails (i.e., on average

Table 3.3 Suggestions for how systematic reviews can deal with predatory journals

Research component	Suggestion number/section
For the review protocol:	1 Detail your methods for addressing the potential for predatory journal articles being captured in your search. (a) Specify how you will determine if an included article meets the criteria for being in a "predatory" journal. (b) Note how you will deal with included articles you determine to be from "predatory" journals. 1 Determine whether included studies are published in open access journals. To do so we suggest the following: (a) If included studies are published in open access (OA) journals, check to determine if the journal is listed in the DOAJ. If yes, presume the journal is legitimate. (b) If included OA journals are not listed in the DOAJ, check to see if the journal is a member of COPE (Committee On Publication Ethics). Note that you should check the COPE membership directory, rather than assume a statement of membership on a journals website is accurate. If yes, presume the journal is legitimate. (c) If included OA journals are not in the DOAJ and not COPE members, review the journal website for characteristics of predatory journals. We suggest that if two or more salient features of predatory journals are present that the journal be classified as predatory. 2 For quantitative analyses, conduct a sensitivity analysis with predatory papers excluded from the synthesis. 3 For qualitative analyses, synthesise results both with and without predatory papers included. 4 Discuss the presence and implications of predatory papers, where relevant.

Source: Rice et al. (2021, p. 3).

158 during the study) and some trainees and assistant professors had more than 30 emails during the study. In another study, McKenzie and colleagues (2021) reported that an academic surgeon received 608 fraudulent phishing emails via his hospital-provided email account over a six-month period.

3.7   Checklists and flow diagrams to identify predatory journals

To help academics identify predatory journals checklists and flow diagrams have been developed (Deora et al., 2021; Richtig et al., 2018). Arguably, Cukier and colleagues (2020) have compiled one of the most comprehensive lists of checklists that can help researchers identify predatory biomedical journals. Most instruments that they listed were published in English ($n = 90$, 97%) and could be completed in less than five minutes ($n = 68$, 73%). Richtig and colleagues (2018) have created a flow diagram that can be used to understand the main distinctions between predatory, subscription-based, and open access journals. As illustrated, the peer review process is one of the main distinctions between predatory journals and other journal types. Unlike open access and subscription-based journals, the articles that are published in predatory journals are not subjected to this process (see Figure 3.9). Using this diagram, scholars can avoid submitting their manuscripts to predatory publishers which will inevitably preserve their academic reputation and the discipline that they study.

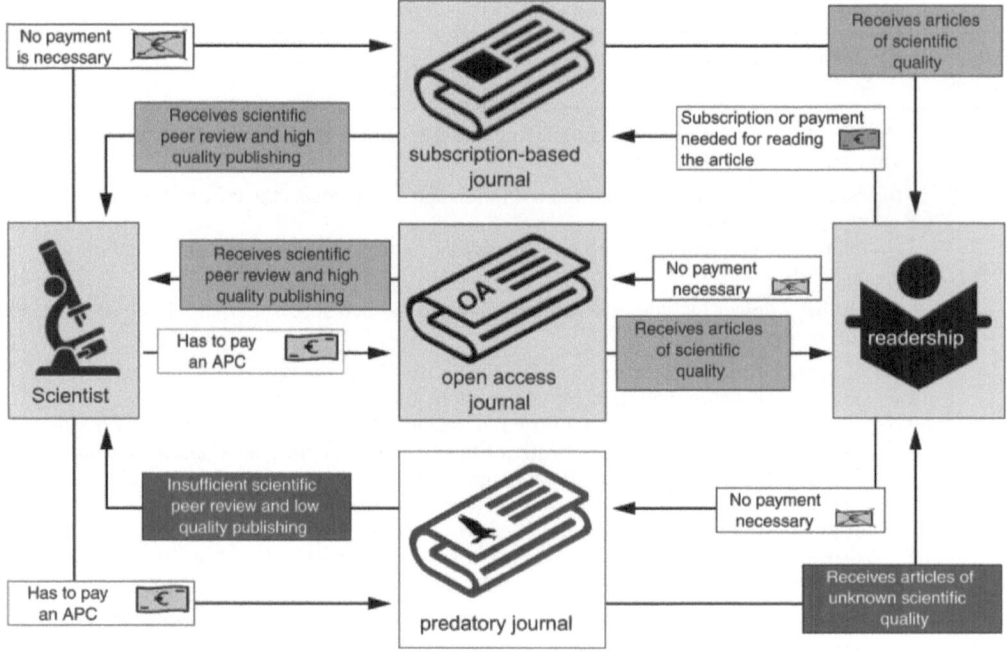

Figure 3.9 The relationships between different types of journals and the author and reader

Source: Richtig et al. (2018, p. 1443).

3.8   Conclusion

In this chapter the main concepts in journalology were presented. It began with a description and historical origins of journalology, followed by a description of the peer review process and the main tasks that peer reviewers typically perform. Several common problems and solutions with this process were outlined. Potential problems within the publication and dissemination process of preprinted articles were described. Finally, the emergence and detrimental impacts of predatory journals were explained. It is hoped that the contents of this chapter have given the reader a more elaborate understanding about journalology, and the processes involved in the dissemination of research.

Additional readings

Ali, M. J. (2022). Manuscript checklist for a scientific publication. *Seminars in Ophthalmology, 37*(1), 1–2. https://doi.org/10.1080/08820538.2022.2021493

Bourne, P. E., Polka, J. K., Vale, R. D., & Kiley, R. (2017). Ten simple rules to consider regarding preprint submission. *PLoS Computational Biology, 13*(5), e1005473. https://doi.org/10.1371/journal.pcbi.1005473

Hamilton, D. G., Fraser, H., Hoekstra, R., & Fidler, F. (2020). Journal policies and editors' opinions on peer review. *eLife, 9*, e62529. https://doi.org/10.7554/eLife.62529

Ho, R. C., Mak, K. K., Tao, R., Lu, Y., Day, J. R., & Pan, F. (2013). Views on the peer review system of biomedical journals: An online survey of academics from high-ranking universities. *BMC Medical Research Methodology, 13*, 74. https://doi.org/10.1186/1471-2288-13-74

Karhulahti, V. M., & Backe, H. J. (2021). Transparency of peer review: A semi-structured interview study with chief editors from social sciences and humanities. *Research Integrity and Peer Review, 6*(1), 13. https://doi.org/10.1186/s41073-021-00116-4

Leonard, M., Stapleton, S., Collins, P., Selfe, T. K., & Cataldo, T. (2021). Ten simple rules for avoiding predatory publishing scams. *PLoS Computational Biology, 17*(9), e1009377. https://doi.org/10.1371/journal.pcbi.1009377

Mahmić-Kaknjo, M., Utrobičić, A., & Marušić, A. (2021). Motivations for performing scholarly prepublication peer review: A scoping review. *Accountability in Research, 28*(5), 297–329. https://doi.org/10. 1080/08989621.2020.1822170

Muscarella, R., & Poorter, L. (2022). Ten simple rules for managing communications with a large number of coauthors. *PLoS Computational Biology, 18*(6), e1010185. https://doi.org/10.1371/journal.pcbi.1010185

Noble, W. S. (2017). Ten simple rules for writing a response to reviewers. *PLoS Computational Biology, 13*(10), e1005730. https://doi.org/10.1371/journal.pcbi.1005730

References

Aczel, B., Szaszi, B., & Holcombe, A. O. (2021). A billion-dollar donation: Estimating the cost of researchers' time spent on peer review. *Research Integrity and Peer Review, 6*(1), 14. https://doi.org/10.1186/ s41073-021-00118-2

Agathokleous, E. (2022). Mastering the scientific peer review process: Tips for young authors from a young senior editor. *Journal of Forestry Research, 33*(1), 1–20. https://doi.org/10.1007/s11676-021-01388-8

Albert, A., Gow, J. L., Cobra, A., & Vines, T. H. (2016). Is it becoming harder to secure reviewers for peer review? A test with data from five ecology journals. *Research Integrity and Peer Review, 1*, 14. https:// doi.org/10.1186/s41073-016-0022-7

American Medical Writers Association, European Medical Writers Association, & International Society for Medical Publication Professionals (2021). AMWA-EMWA-ISMPP joint position statement on medical publications, preprints, and peer review. *Current Medical Research and Opinion, 37*(5), 861–866. https://doi.org/10.1080/03007995.2021.1900365

Annesley, T. M. (2011). Top 10 tips for responding to reviewer and editor comments. *Clinical Chemistry, 57*(4), 551–554. https://doi.org/10.1373/clinchem.2011.162388

Asiimwe, I. G., & Rumona, D. (2016). Publication proportions for registered breast cancer trials: Before and following the introduction of the ClinicalTrials.gov results database. *Research Integrity and Peer Review, 1*, 10. https://doi.org/10.1186/s41073-016-0017-4

Ayorinde, A. A., Williams, I., Mannion, R., Song, F., Skrybant, M., Lilford, R. J., & Chen, Y. F. (2020). Publication and related biases in health services research: A systematic review of empirical evidence. *BMC Medical Research Methodology, 20*(1), 137. https://doi.org/10.1186/s12874-020-01010-1

Baker, W. L., DiDomenico, R. J., & Haines, S. T. (2017). Improving peer review: What authors can do. *American Journal of Health-system Pharmacy, 74*(24), 2076–2079. https://doi.org/10.2146/ajhp170187

Barnett, A. G., Campbell, M. J., Shield, C., Farrington, A., Hall, L., Page, K., Gardner, A., Mitchell, B. G., & Graves, N. (2016). The high costs of getting ethical and site-specific approvals for multi-centre research. *Research Integrity and Peer Review, 1*, 16. https://doi.org/10.1186/s41073-016-0023-6

Barnett, A. G., Clarke, P., Vaquette, C., & Graves, N. (2017). Using democracy to award research funding: An observational study. *Research Integrity and Peer Review, 2*, 16. https://doi.org/10.1186/ s41073-017-0040-0

Beall, J. (2016a). Dangerous predatory publishers threaten medical research. *Journal of Korean Medical Science, 31*(10), 1511–1513. https://doi.org/10.3346/jkms.2016.31.10.1511

Beall, J. (2016b). Best practices for scholarly authors in the age of predatory journals. *Annals of the Royal College of Surgeons of England, 98*(2), 77–79. https://doi.org/10.1308/rcsann.2016.0056

Beall, J. (2017). What I learned from predatory publishers. *Biochemia Medica, 27*(2), 273–278. https://doi. org/10.11613/BM.2017.029

Bennett, M., & Goodall, E. (2022). *Addressing Underserved Populations in Autism Spectrum Research*. Emerald Publishing Limited, Bingley.

Bornmann, L., & Daniel, H. D. (2010). Do author-suggested reviewers rate submissions more favorably than editor-suggested reviewers? A study on atmospheric chemistry and physics. *PLoS One, 5*(10), e13345. https://doi.org/10.1371/journal.pone.0013345

Boughton, S. L., Kowalczuk, M. K., Meerpohl, J. J., Wager, E., & Moylan, E. C. (2018). Research Integrity and Peer Review-past highlights and future directions. *Research Integrity and Peer Review, 3*, 3. https:// doi.org/10.1186/s41073-018-0047-1

Bouter, L. M., Tijdink, J., Axelsen, N., Martinson, B. C., & Ter Riet, G. (2016). Ranking major and minor research misbehaviors: Results from a survey among participants of four World Conferences on Research Integrity. *Research Integrity and Peer Review, 1*, 17. https://doi.org/10.1186/s41073-016-0024-5

Boyer, S., Ikeda, T., Lefort, M. C., Malumbres-Olarte, J., & Schmidt, J. M. (2017). Percentage-based author contribution index: A universal measure of author contribution to scientific articles. *Research Integrity and Peer Review, 2*, 18. https://doi.org/10.1186/s41073-017-0042-y

Bozzo, A., Bali, K., Evaniew, N., & Ghert, M. (2017). Retractions in cancer research: A systematic survey. *Research Integrity and Peer Review, 2*, 5. https://doi.org/10.1186/s41073-017-0031-1

Button, K. S., Bal, L., Clark, A., & Shipley, T. (2016). Preventing the ends from justifying the means: Withholding results to address publication bias in peer-review. *BMC Psychology, 4*(1), 59. https://doi.org/10.1186/s40359-016-0167-7

Byrne, J. A. (2016). Improving the peer review of narrative literature reviews. *Research Integrity and Peer Review, 1*, 12. https://doi.org/10.1186/s41073-016-0019-2

Chambers, C. (2014). *Registered reports: A step change in scientific publishing.* https://www.elsevier.com/reviewers-update/story/innovation-in-publishing/registered-reports-a-step-change-in-scientific-publishing

Cosgrove, A., & Cheifet, B. (2018). Transparent peer review trial: The results. *Genome Biology, 19*(1), 206. https://doi.org/10.1186/s13059-018-1584-0

Coveney, J., Herbert, D. L., Hill, K., Mow, K. E., Graves, N., & Barnett, A. (2017). 'Are you siding with a personality or the grant proposal?': Observations on how peer review panels function. *Research Integrity and Peer Review, 2*, 19. https://doi.org/10.1186/s41073-017-0043-x

Crijns, T. J., Ottenhoff, J., & Ring, D. (2021). The effect of peer review on the improvement of rejected manuscripts. *Accountability in Research, 28*(8), 517–527. https://doi.org/10.1080/08989621.2020.1869547

Cukier, S., Helal, L., Rice, D. B., Pupkaite, J., Ahmadzai, N., Wilson, M., Skidmore, B., Lalu, M. M., & Moher, D. (2020). Checklists to detect potential predatory biomedical journals: A systematic review. *BMC Medicine, 18*(1), 104. https://doi.org/10.1186/s12916-020-01566-1

Dennehy, J., Hoxie, I., di Schiavi, E., & Onorato, G. (2021). Reviewing as a career milestone: A discussion on the importance of including trainees in the peer review process. *Communications Biology, 4*(1), 1126. https://doi.org/10.1038/s42003-021-02645-6

Deora, H., Tripathi, M., Chaurasia, B., & Grotenhuis, J. A. (2021). Avoiding predatory publishing for early career neurosurgeons: What should you know before you submit? *Acta Neurochirurgica, 163*(1), 1–8. https://doi.org/10.1007/s00701-020-04546-9 .

Dinakaran, D., Anaka, M., & Mackey, J. R. (2021). Proposal for 'segmented peer review' of multidisciplinary papers. *Translational Oncology, 14*(2), 100985. https://doi.org/10.1016/j.tranon.2020.100985

Dunn, A. G., Coiera, E., Mandl, K. D., & Bourgeois, F. T. (2016). Conflict of interest disclosure in biomedical research: A review of current practices, biases, and the role of public registries in improving transparency. *Research Integrity and Peer Review, 1*, 1. https://doi.org/10.1186/s41073-016-0006-7

Duyx, B., Urlings, M., Swaen, G., Bouter, L. M., & Zeegers, M. P. (2017). Selective citation in the literature on swimming in chlorinated water and childhood asthma: A network analysis. *Research Integrity and Peer Review, 2*, 17. https://doi.org/10.1186/s41073-017-0041-z

Elmore, S. A. (2018). Preprints: What role do these have in communicating scientific results? *Toxicologic Pathology, 46*(4), 364–365. https://doi.org/10.1177/0192623318767322

Fox, C. W., Albert, A., & Vines, T. H. (2017). Recruitment of reviewers is becoming harder at some journals: A test of the influence of reviewer fatigue at six journals in ecology and evolution. *Research Integrity and Peer Review, 2*, 3. https://doi.org/10.1186/s41073-017-0027-x

Galica, J., Chee-A-Tow, A., Gupta, S., Jaiswal, A., Monsour, A., Tricco, A. C., Cobey, K. D., & Butcher, N. J. (2018). Learning best-practices in journalology: Course description and attendee insights into the inaugural EQUATOR Canada Publication School. *BMC Proceedings, 12*(Suppl 10), 18. https://doi.org/10.1186/s12919-018-0155-4

Gerwing, T. G., Allen Gerwing, A. M., Avery-Gomm, S., Choi, C. Y., Clements, J. C., & Rash, J. A. (2020). Quantifying professionalism in peer review. *Research Integrity and Peer Review, 5*, 9. https://doi.org/10.1186/s41073-020-00096-x

Glonti, K., Cauchi, D., Cobo, E., Boutron, I., Moher, D., & Hren, D. (2019). A scoping review on the roles and tasks of peer reviewers in the manuscript review process in biomedical journals. *BMC Medicine, 17*(1), 118. https://doi.org/10.1186/s12916-019-1347-0

Gray, R., Badnapurkar, A., Hassanein, E., Thomas, D., Barguir, L., Baker, C., Jones, M., Bressington, D., Brown, E., & Topping, A. (2017). Registration of randomized controlled trials in nursing journals. *Research Integrity and Peer Review, 2*, 8. https://doi.org/10.1186/s41073-017-0036-9

Hamilton, S., Bernstein, A. B., Blakey, G., Fagan, V., Farrow, T., Jordan, D., Seiler, W., Shannon, A., Gertel, A., & Budapest Working Group. (2016). Developing the clarity and openness in reporting: E3-based (CORE) Reference user manual for creation of clinical study reports in the era of clinical trial transparency. *Research Integrity and Peer Review, 1*, 4. https://doi.org/10.1186/s41073-016-0009-4

Haneef, R., Ravaud, P., Baron, G., Ghosn, L., & Boutron, I. (2017). Factors associated with online media attention to research: A cohort study of articles evaluating cancer treatments. *Research Integrity and Peer Review, 2*, 9. https://doi.org/10.1186/s41073-017-0033-z

Harriman, S. L., Kowalczuk, M. K., Simera, I., & Wager, E. (2016). A new forum for research on research integrity and peer review. *Research Integrity and Peer Review, 1*, 5. https://doi.org/10.1186/s41073-016-0010-y

Heidari, S., Babor, T. F., De Castro, P., Tort, S., & Curno, M. (2016). Sex and gender equity in research: Rationale for the SAGER guidelines and recommended use. *Research Integrity and Peer Review, 1*, 2. https://doi.org/10.1186/s41073-016-0007-6

Heim, A., Ravaud, P., Baron, G., & Boutron, I. (2018). Designs of trials assessing interventions to improve the peer review process: A vignette-based survey. *BMC Medicine, 16*(1), 191. https://doi.org/10.1186/s12916-018-1167-7

Higgins, J. R., Lin, F. C., & Evans, J. P. (2016). Plagiarism in submitted manuscripts: Incidence, characteristics and optimization of screening-case study in a major specialty medical journal. *Research Integrity and Peer Review, 1*, 13. https://doi.org/10.1186/s41073-016-0021-8

Hoffman, A. J. (2022). A modest proposal to the peer review process: A collaborative and interdisciplinary approach in the assessment of scholarly communication. *Research Ethics, 18*(1), 84–91. https://doi.org/10.1177/17470161211051230

Hrynaszkiewicz, I., Khodiyar, V., Hufton, A. L., & Sansone, S. A. (2016). Publishing descriptions of non-public clinical datasets: Proposed guidance for researchers, repositories, editors and funding organisations. *Research Integrity and Peer Review, 1*, 6. https://doi.org/10.1186/s41073-016-0015-6

Klebel, T., Reichmann, S., Polka, J., McDowell, G., Penfold, N., Hindle, S., & Ross-Hellauer, T. (2020). Peer review and preprint policies are unclear at most major journals. *PLoS One, 15*(10), e0239518. https://doi.org/10.1371/journal.pone.0239518

Korevaar, D. A., Cohen, J. F., Reitsma, J. B., Bruns, D. E., Gatsonis, C. A., Glasziou, P. P., Irwig, L., Moher, D., de Vet, H., Altman, D. G., Hooft, L., & Bossuyt, P. (2016). Updating standards for reporting diagnostic accuracy: The development of STARD 2015. *Research Integrity and Peer Review, 1*, 7. https://doi.org/10.1186/s41073-016-0014-7

Kovanis, M., Porcher, R., Ravaud, P., & Trinquart, L. (2016). The global burden of journal peer review in the biomedical literature: Strong imbalance in the collective enterprise. *PLoS One, 11*(11), e0166387. https://doi.org/10.1371/journal.pone.0166387

Kowalczuk, M. K., Dudbridge, F., Nanda, S., Harriman, S. L., Patel, J., & Moylan, E. C. (2015). Retrospective analysis of the quality of reports by author-suggested and non-author-suggested reviewers in journals operating on open or single-blind peer review models. *BMJ Open, 5*(9), e008707. https://doi.org/10.1136/bmjopen-2015-008707

Krasowski, M. D., Lawrence, J. C., Briggs, A. S., & Ford, B. A. (2019). Burden and characteristics of unsolicited emails from medical/scientific journals, conferences, and webinars to faculty and trainees at an academic pathology department. *Journal of Pathology Informatics, 10*, 16. https://doi.org/10.4103/jpi.jpi_12_19

Kretser, A., Murphy, D., Bertuzzi, S., Abraham, T., Allison, D. B., Boor, K. J., Dwyer, J., Grantham, A., Harris, L. J., Hollander, R., Jacobs-Young, C., Rovito, S., Vafiadis, D., Woteki, C., Wyndham, J., & Yada, R. (2019). Scientific integrity principles and best practices: Recommendations from a scientific integrity consortium. *Science and Engineering Ethics, 25*(2), 327–355. https://doi.org/10.1007/s11948-019-00094-3

Krishan, K., & Kanchan, T. (2019). Open centres for Journalology in Universities and Institutions. *Science and Engineering Ethics, 25*(4), 1259–1260. https://doi.org/10.1007/s11948-018-0047-z

LeBlanc, A. G., Barnes, J. D., Saunders, T. J., Tremblay, M. S., & Chaput, J. P. (2023). Scientific sinkhole: Estimating the cost of peer review based on survey data with snowball sampling. *Research Integrity and Peer Review, 8*(1), 3. https://doi.org/10.1186/s41073-023-00128-2

Liang, Y. (2018). Should authors suggest reviewers? A comparative study of the performance of author-suggested and editor-selected reviewers at a biological journal. *Learned Publishing, 31*(3), 216–221. https://doi.org/10.1002/leap.1166

McKenzie, M., Nickerson, D., & Ball, C. G. (2021). Predatory publishing solicitation: A review of a single surgeon's inbox and implications for information technology resources at an organizational level. *Canadian Journal of Surgery, 64*(3), E351–E357. https://doi.org/10.1503/cjs.003020

Moskovitz, C. (2017). Text recycling in health sciences research literature: A rhetorical perspective. *Research Integrity and Peer Review, 2*, 1. https://doi.org/10.1186/s41073-017-0025-z

Nature Communications. (2016). Transparent peer review one year on. *Nature Communications, 7*, 13626. https://doi.org/10.1038/ncomms13626

O'Brien, S. P., Chan, D., Leung, F., Ko, E. J., Kwak, J. S., Gwon, T., Lee, J. M., Lee, M., Nolte, H., Gommel, M., Sponholz, G., Krastev, Y., Sandiran, Y., Connell, J., Solomon, N., Krasovec, U. O., Sribar, R., Martinson, B. C., Thrush, C. R., ... Bouter, L. (2016). Proceedings of the 4th world conference. *Research Integrity and Peer Review* (Suppl 1), 9. https://doi.org/10.1186/s41073-016-0012-9

Page, S. A., & Nyeboer, J. (2017). Improving the process of research ethics review. *Research Integrity and Peer Review, 2*, 14. https://doi.org/10.1186/s41073-017-0038-7

Patel, J., Pierce, M., Boughton, S. L., & Baldeweg, S. E. (2017). Do peer review models affect clinicians' trust in journals? A survey of junior doctors. *Research Integrity and Peer Review, 2*, 11. https://doi.org/10.1186/s41073-017-0029-8

Räsänen, L., & Moore, E. (2016). Critical evaluation of the guidelines of the Finnish Advisory Board on Research Integrity and of their application. *Research Integrity and Peer Review, 1*, 15. https://doi.org/10.1186/s41073-016-0020-9

Ravinetto, R., Caillet, C., Zaman, M. H., Singh, J. A., Guerin, P. J., Ahmad, A., Durán, C. E., Jesani, A., Palmero, A., Merson, L., Horby, P. W., Bottieau, E., Hoffmann, T., & Newton, P. N. (2021). Preprints in times of COVID19: The time is ripe for agreeing on terminology and good practices. *BMC Medical Ethics, 22*(1), 106. https://doi.org/10.1186/s12910-021-00667-7

Rice, D. B., Skidmore, B., & Cobey, K. D. (2021). Dealing with predatory journal articles captured in systematic reviews. *Systematic Reviews, 10*(1), 175. https://doi.org/10.1186/s13643-021-01733-2

Richtig, G., Berger, M., Lange-Asschenfeldt, B., Aberer, W., & Richtig, E. (2018). Problems and challenges of predatory journals. *Journal of the European Academy of Dermatology and Venereology, 32*(9), 1441–1449. https://doi.org/10.1111/jdv.15039

Roig, M. (2017). Comment on Cary Moskovitz' "Text recycling in health sciences literature: A rhetorical perspective". *Research Integrity and Peer Review, 2*, 2. https://doi.org/10.1186/s41073-017-0026-y

Ross-Hellauer, T., Deppe, A., & Schmidt, B. (2017). Survey on open peer review: Attitudes and experience amongst editors, authors and reviewers. *PLoS One, 12*(12), e0189311. https://doi.org/10.1371/journal.pone.0189311

Rowhani-Farid, A., Allen, M., & Barnett, A. G. (2017). What incentives increase data sharing in health and medical research? A systematic review. *Research Integrity and Peer Review, 2*, 4. https://doi.org/10.1186/s41073-017-0028-9

Scarrow, G., Angus, D., & Holmes, B. J. (2017). Reviewer training to assess knowledge translation in funding applications is long overdue. *Research Integrity and Peer Review, 2*, 13. https://doi.org/10.1186/s41073-017-0037-8

Schmidt, B., Ross-Hellauer, T., van Edig, X., & Moylan, E. C. (2018). Ten considerations for open peer review. *F1000Research, 7*, 969. https://doi.org/10.12688/f1000research.15334.1

Shanahan, D. R., Lopes de Sousa, I., & Marshall, D. M. (2017). Simple decision-tree tool to facilitate author identification of reporting guidelines during submission: A before-after study. *Research Integrity and Peer Review, 2*, 20. https://doi.org/10.1186/s41073-017-0044-9

Shopovski, J., Bolek, C., & Bolek, M. (2020). Characteristics of peer review reports: Editor-suggested versus author-suggested reviewers. *Science and Engineering Ethics, 26*(2), 709–726. https://doi.org/10.1007/s11948-019-00118-y

Silbiger, N. J., & Stubler, A. D. (2019). Unprofessional peer reviews disproportionately harm underrepresented groups in STEM. *PeerJ, 7*, e8247. https://doi.org/10.7717/peerj.8247

Smith, E. M. (2021). Reimagining the peer-review system for translational health science journals. *Clinical and Translational Science, 14*(4), 1210–1221. https://doi.org/10.1111/cts.13050

Sousa, F., Nadanovsky, P., Dhyppolito, I. M., & Santos, A. (2021). One year of unsolicited e-mails: The modus operandi of predatory journals and publishers. *Journal of Dentistry, 109*, 103618. https://doi.org/10.1016/j.jdent.2021.103618

Struthers, C., Harwood, J., de Beyer, J. A., Dhiman, P., Logullo, P., & Schlüssel, M. (2021). GoodReports: Developing a website to help health researchers find and use reporting guidelines. *BMC Medical Research Methodology, 21*(1), 217. https://doi.org/10.1186/s12874-021-01402-x

Tennant, J. P., Crane, H., Crick, T., Davila, J., Enkhbayar, A., Havemann, J., Kramer, B., Martin, R., Masuzzo, P., Nobes, A., Rice, C., Rivera-López, B., Ross-Hellauer, T., Sattler, S., Thacker, P. D., & Vanholsbeeck, M. (2019). Ten hot topics around scholarly publishing. *Publications, 7*(2), 34. https://doi.org/10.3390/publications7020034

Tennant, J. P., & Ross-Hellauer, T. (2020). The limitations to our understanding of peer review. *Research Integrity and Peer Review, 5*, 6. https://doi.org/10.1186/s41073-020-00092-1

Tomkins, A., Zhang, M., & Heavlin, W. D. (2017). Reviewer bias in single- versus double-blind peer review. *Proceedings of the National Academy of Sciences of the United States of America, 114*(48), 12708–12713. https://doi.org/10.1073/pnas.1707323114

Tsuyuki, R. T., Al Hamarneh, Y. N., Bermingham, M., Duong, E., Okada, H., & Beall, J. (2017). Predatory publishers: Implications for pharmacy practice and practitioners. *Canadian Pharmacists Journal, 150*(5), 274–275. https://doi.org/10.1177/1715163517725269

van der Vet, P. E., & Nijveen, H. (2016). Propagation of errors in citation networks: A study involving the entire citation network of a widely cited paper published in, and later retracted from, the journal Nature. *Research Integrity and Peer Review, 1*, 3. https://doi.org/10.1186/s41073-016-0008-5

Vasconcelos, S., Watanabe, E., Garcia, L. P., Duarte, E., Cassimiro, M. C., Diós-Borges, M. M. P., Soares, A. M. M., Debenedito Silva, C. H., Santa Rosa, A. A. P., Fófano, G. A., Pinheiro, H. S., Gollner, A. M., Santos, C. C., Vasconcelos, S. M. R., Machado, D. C., Souza, P. V. S., Souza, R. T., Ribeiro, M. D., Vasconcelos, S. M. R., … Abreu, A. S. C. (2017). Proceedings from the IV Brazilian Meeting on Research Integrity, Science and Publication Ethics (IV BRISPE). *Research Integrity and Peer Review, 2* (Suppl 1):12. https://doi.org/10.1186/s41073-017-0035-x

Vaught, M., Jordan, D. C., & Bastian, H. (2017). Concern noted: A descriptive study of editorial expressions of concern in PubMed and PubMed Central. *Research Integrity and Peer Review, 2*, 10. https://doi.org/10.1186/s41073-017-0030-2

Wager, E., Altman, D. G., Simera, I., & Toma, T. P. (2016). Do declarative titles affect readers' perceptions of research findings? A randomized trial. *Research Integrity and Peer Review, 1*, 11. https://doi.org/10.1186/s41073-016-0018-3

Wallach, J. D., Egilman, A. C., Gopal, A. D., Swami, N., Krumholz, H. M., & Ross, J. S. (2018). Biomedical journal speed and efficiency: A cross-sectional pilot survey of author experiences. *Research Integrity and Peer Review, 3*, 1. https://doi.org/10.1186/s41073-017-0045-8

Welch, V., Doull, M., Yoganathan, M., Jull, J., Boscoe, M., Coen, S. E., Marshall, Z., Pardo, J. P., Pederson, A., Petkovic, J., Puil, L., Quinlan, L., Shea, B., Rader, T., Runnels, V., & Tudiver, S. (2017). Reporting of sex and gender in randomized controlled trials in Canada: A cross-sectional methods study. *Research Integrity and Peer Review, 2*, 15. https://doi.org/10.1186/s41073-017-0039-6

Wong, V., Strowd, R. E., 3rd, Aragón-García, R., Moon, Y. P., Ford, B., Haut, S. R., Kass, J. S., London, Z. N., Mays, M., Milligan, T. A., Price, R. S., Reynolds, P. S., Selwa, L. M., Spencer, D. C., & Elkind, M. (2017). Mentored peer review of standardized manuscripts as a teaching tool for residents: A pilot randomized controlled multi-center study. *Research Integrity and Peer Review, 2*, 6. https://doi.org/10.1186/s41073-017-0032-0

Wood, K. E., & Krasowski, M. D. (2020). Academic e-mail overload and the burden of "academic spam". *Academic Pathology, 7*, 2374289519898858. https://doi.org/10.1177/2374289519898858

Appendix 3.1

Checklist for analytical cross-sectional studies

Introduction

JBI is an international research organisation based in the Faculty of Health and Medical Sciences at the University of Adelaide, South Australia. JBI develops and delivers unique evidence-based information, software, education and training designed to improve healthcare practice and health outcomes. With over 70 Collaborating Entities, servicing over 90 countries, JBI is a recognised global leader in evidence-based healthcare.

JBI systematic reviews

The core of evidence synthesis is the systematic review of literature of a particular intervention, condition or issue. The systematic review is essentially an analysis of the available literature (i.e., evidence) and a judgment of the effectiveness or otherwise of a practice, involving a series of complex steps. JBI takes a particular view on what counts as evidence and the methods utilised to synthesise those different types of evidence. In line with this broader view of evidence, JBI has developed theories, methodologies and rigorous processes for the critical appraisal and synthesis of these diverse forms of evidence in order to aid in clinical decision-making in healthcare. There now exists JBI guidance for conducting reviews of effectiveness research, qualitative research, prevalence/incidence, aetiology/risk, economic evaluations, text/opinion, diagnostic test accuracy, mixed-methods, umbrella reviews and scoping reviews. Further information regarding JBI systematic reviews can be found in the JBI Evidence Synthesis Manual.

JBI critical appraisal tools

All systematic reviews incorporate a process of critique or appraisal of the research evidence. The purpose of this appraisal is to assess the methodological quality of a study and to determine the extent to which a study has addressed the possibility of bias in its design, conduct and analysis. All papers selected for inclusion in the systematic review (i.e., those that meet the inclusion criteria described in the protocol) need to be subjected to rigorous appraisal by two critical appraisers. The results of this appraisal can then be used to inform synthesis and interpretation of the results of the study. JBI critical appraisal tools have been developed by the JBI and collaborators and approved by the JBI Scientific Committee following extensive peer review. Although designed for use in systematic reviews, JBI critical appraisal tools can also be used when creating Critically Appraised Topics (CAT), in journal clubs and as an educational tool.

JBI critical appraisal checklist for analytical cross-sectional studies

Reviewer_____ Date _____

Author_____ Year _____ Record Number _____

	Yes	No	Unclear	Not applicable
1 Were the criteria for inclusion in the sample clearly defined?	□	□	□	□
2 Were the study subjects and the setting described in detail?	□	□	□	□
3 Was the exposure measured in a valid and reliable way?	□	□	□	□
4 Were objective, standard criteria used for measurement of the condition?	□	□	□	□
5 Were confounding factors identified?	□	□	□	□
6 Were strategies to deal with confounding factors stated?	□	□	□	□
7 Were the outcomes measured in a valid and reliable way?	□	□	□	□
8 Was appropriate statistical analysis used?	□	□	□	□

Overall appraisal: Include □   Exclude □   Seek further info □

Comments (Including reason for exclusion)

_____

_____

_____

Explanation of analytical cross-sectional studies critical appraisal

How to cite: Moola S, Munn Z, Tufanaru C, Aromataris E, Sears K, Sfetcu R, Currie M, Qureshi R, Mattis P, Lisy K, Mu P-F. Chapter 7: Systematic reviews of etiology and risk. In: Aromataris E, Munn Z (Editors). *JBI Manual for Evidence Synthesis.* JBI, 2020. Available from https://synthesismanual.jbi.global

Analytical cross-sectional studies critical appraisal tool

Answers: Yes, No, Unclear or Not/Applicable

1 Were the criteria for inclusion in the sample clearly defined?
The authors should provide clear inclusion and exclusion criteria that they developed prior to recruitment of the study participants. The inclusion/exclusion criteria should be specified (e.g., risk and stage of disease progression) with sufficient detail and all the necessary information critical to the study.

2 Were the study subjects and the setting described in detail?
The study sample should be described in sufficient detail so that other researchers can determine if it is comparable to the population of interest to them. The authors should provide

a clear description of the population from which the study participants were selected or recruited, including demographics, location, and time period.

3 Was the exposure measured in a valid and reliable way?

The study should clearly describe the method of measurement of exposure. Assessing validity requires that a 'gold standard' is available to which the measure can be compared. The validity of exposure measurement usually relates to whether a current measure is appropriate or whether a measure of past exposure is needed.

Reliability refers to the processes included in an epidemiological study to check the repeatability of measurements of the exposures. These usually include intra-observer reliability and inter-observer reliability.

4 Were objective, standard criteria used for measurement of the condition?

It is useful to determine if patients were included in the study based on either a specified diagnosis or definition. This is more likely to decrease the risk of bias. Characteristics are another useful approach to matching groups, and studies that did not use specified diagnostic methods or definitions should provide evidence on matching by key characteristics.

5 Were confounding factors identified?

Confounding has occurred where the estimated intervention exposure effect is biased by the presence of some difference between the comparison groups (apart from the exposure investigated/of interest). Typical confounders include baseline characteristics, prognostic factors, or concomitant exposures (e.g. smoking). A confounder is a difference between the comparison groups and it influences the direction of the study results. A high-quality study at the level of cohort design will identify the potential confounders and measure them (where possible). This is difficult for studies where behavioural, attitudinal or lifestyle factors may impact on the results.

6 Were strategies to deal with confounding factors stated?

Strategies to deal with effects of confounding factors may be dealt within the study design or in data analysis. By matching or stratifying sampling of participants, effects of confounding factors can be adjusted for. When dealing with adjustment in data analysis, assess the statistics used in the study. Most will be some form of multivariate regression analysis to account for the confounding factors measured.

7 Were the outcomes measured in a valid and reliable way?

Read the methods section of the paper. If, for example, lung cancer is assessed based on existing definitions or diagnostic criteria, then the answer to this question is likely to be yes. If lung cancer is assessed using observer reported, or self-reported scales, the risk of over- or under-reporting is increased, and objectivity is compromised. Importantly, determine if the measurement tools used were validated instruments as this has a significant impact on outcome assessment validity.

Having established the objectivity of the outcome measurement (e.g. lung cancer) instrument, it's important to establish how the measurement was conducted. Were those involved in collecting data trained or educated in the use of the instrument/s? (e.g. radiographers). If there was more than one data collector, were they similar in terms of level of education, clinical or research experience, or level of responsibility in the piece of research being appraised?

8 **Was appropriate statistical analysis used?**

As with any consideration of statistical analysis, consideration should be given to whether there was a more appropriate alternate statistical method that could have been used. The methods section should be detailed enough for reviewers to identify which analytical techniques were used (in particular, regression or stratification) and how specific confounders were measured.

For studies utilising regression analysis, it is useful to identify if the study identified which variables were included and how they related to the outcome. If stratification was the analytical approach used, were the strata of analysis defined by the specified variables? Additionally, it is also important to assess the appropriateness of the analytical strategy in terms of the assumptions associated with the approach as differing methods of analysis are based on differing assumptions about the data and how it will respond.

Checklist for case control studies

Introduction

JBI is an international research organisation based in the Faculty of Health and Medical Sciences at the University of Adelaide, South Australia. JBI develops and delivers unique evidence-based information, software, education and training designed to improve healthcare practice and health outcomes. With over 70 Collaborating Entities, servicing over 90 countries, JBI is a recognised global leader in evidence-based healthcare.

JBI systematic reviews

The core of evidence synthesis is the systematic review of literature of a particular intervention, condition or issue. The systematic review is essentially an analysis of the available literature (i.e., evidence) and a judgment of the effectiveness or otherwise of a practice, involving a series of complex steps. JBI takes a particular view on what counts as evidence and the methods utilised to synthesise those different types of evidence. In line with this broader view of evidence, JBI has developed theories, methodologies and rigorous processes for the critical appraisal and synthesis of these diverse forms of evidence in order to aid in clinical decision-making in healthcare. There now exists JBI guidance for conducting reviews of effectiveness research, qualitative research, prevalence/incidence, aetiology/risk, economic evaluations, text/opinion, diagnostic test accuracy, mixed-methods, umbrella reviews and scoping reviews. Further information regarding JBI systematic reviews can be found in the JBI Evidence Synthesis Manual.

JBI critical appraisal tools

All systematic reviews incorporate a process of critique or appraisal of the research evidence. The purpose of this appraisal is to assess the methodological quality of a study and to determine the extent to which a study has addressed the possibility of bias in its design, conduct and analysis. All papers selected for inclusion in the systematic review (i.e., those that meet the inclusion criteria described in the protocol) need to be subjected to rigorous appraisal by two critical appraisers. The results of this appraisal can then be used to inform synthesis and interpretation of the results of the study. JBI critical appraisal tools have been developed by the JBI and collaborators and approved by the JBI Scientific Committee following extensive peer review. Although designed for use in systematic reviews, JBI critical appraisal tools can also be used when creating Critically Appraised Topics (CAT), in journal clubs and as an educational tool.

JBI critical appraisal checklist for case control studies

Reviewer _____ Date _____

Author _____ Year _____ Record Number _____

	Yes	No	Unclear	Not applicable
1 Were the groups comparable other than the presence of disease in cases or the absence of disease in controls?	□	□	□	□
2 Were cases and controls matched appropriately?	□	□	□	□
3 Were the same criteria used for identification of cases and controls?	□	□	□	□
4 Was exposure measured in a standard, valid and reliable way?	□	□	□	□
5 Was exposure measured in the same way for cases and controls?	□	□	□	□
6 Were confounding factors identified?	□	□	□	□
7 Were strategies to deal with confounding factors stated?	□	□	□	□
8 Were outcomes assessed in a standard, valid and reliable way for cases and controls?	□	□	□	□
9 Was the exposure period of interest long enough to be meaningful?	□	□	□	□
10 Was appropriate statistical analysis used?	□	□	□	□

Overall appraisal:　　　Include　□　　Exclude　□　　Seek further info　□

Comments (Including reason for exclusion)

_____

_____

_____

Explanation of case control studies critical appraisal

How to cite: *Moola S, Munn Z, Tufanaru C, Aromataris E, Sears K, Sfetcu R, Currie M, Qureshi R, Mattis P, Lisy K, Mu P-F. Chapter 7: Systematic reviews of etiology and risk. In: Aromataris E, Munn Z (Editors). JBI Manual for Evidence Synthesis.* JBI, 2020. Available from https://synthesismanual.jbi.global

Case–control studies critical appraisal tool

Answers: Yes, No, Unclear or Not/Applicable

1 **Were the groups comparable other than presence of disease in cases or absence of disease in controls?**

The control group should be representative of the source population that produced the cases. This is usually done by individual matching; wherein controls are selected for each case on the basis of similarity with respect to certain characteristics other than the exposure of interest. Frequency or group matching is an alternative method. Selection bias may result if the groups are not comparable.

2 Were cases and controls matched appropriately?

As in item 1, the study should include clear definitions of the source population. Sources from which cases and controls were recruited should be carefully looked at. For example, cancer registries may be used to recruit participants in a study examining risk factors for lung cancer, which typify population-based case control studies. Study participants may be selected from the target population, the source population, or from a pool of eligible participants (such as in hospital-based case control studies).

3 Were the same criteria used for identification of cases and controls?

It is useful to determine if patients were included in the study based on either a specified diagnosis or definition. This is more likely to decrease the risk of bias. Characteristics are another useful approach to matching groups, and studies that did not use specified diagnostic methods or definitions should provide evidence on matching by key characteristics. A case should be defined clearly. It is also important that controls must fulfil all the eligibility criteria defined for the cases except for those relating to diagnosis of the disease.

4 Was exposure measured in a standard, valid and reliable way?

The study should clearly describe the method of measurement of exposure. Assessing validity requires that a 'gold standard' is available to which the measure can be compared. The validity of exposure measurement usually relates to whether a current measure is appropriate or whether a measure of past exposure is needed.

Case control studies may investigate many different 'exposures' that may or may not be associated with the condition. In these cases, reviewers should use the main exposure of interest for their review to answer this question when using this tool at the study level.

Reliability refers to the processes included in an epidemiological study to check repeatability of measurements of the exposures. These usually include intra-observer reliability and inter-observer reliability.

5 Was exposure measured in the same way for cases and controls?

As in item 4, the study should clearly describe the method of measurement of exposure. The exposure measures should be clearly defined and described in detail. Assessment of exposure or risk factors should have been carried out according to same procedures or protocols for both cases and controls.

6 Were confounding factors identified?

Confounding has occurred where the estimated intervention exposure effect is biased by the presence of some difference between the comparison groups (apart from the exposure investigated/of interest). Typical confounders include baseline characteristics, prognostic factors, or concomitant exposures (e.g. smoking). A confounder is a difference between the comparison groups and it influences the direction of the study results. A high-quality study at the level of case–control design will identify the potential confounders and measure them (where possible). This is difficult for studies where behavioural, attitudinal or lifestyle factors may impact on the results.

7 Were strategies to deal with confounding factors stated?

Strategies to deal with effects of confounding factors may be dealt within the study design or in data analysis. By matching or stratifying sampling of participants, effects of confounding factors can be adjusted for. When dealing with adjustment in data analysis, assess the statistics used in the study. Most will be some form of multivariate regression analysis to account for the confounding factors measured. Look out for a description of statistical methods as regression methods such as logistic regression are usually employed to deal with confounding factors/ variables of interest.

8 **Were outcomes assessed in a standard, valid and reliable way for cases and controls?**
Read the methods section of the paper. If, for example, lung cancer is assessed based on existing definitions or diagnostic criteria, then the answer to this question is likely to be yes. If lung cancer is assessed using observer reported, or self-reported scales, the risk of over- or under-reporting is increased, and objectivity is compromised. Importantly, determine if the measurement tools used were validated instruments as this has a significant impact on outcome assessment validity.

Having established the objectivity of the outcome measurement (e.g. lung cancer) instrument, it's important to establish how the measurement was conducted. Were those involved in collecting data trained or educated in the use of the instrument/s? (e.g. radiographers). If there was more than one data collector, were they similar in terms of level of education, clinical or research experience, or level of responsibility in the piece of research being appraised?

9 **Was the exposure period of interest long enough to be meaningful?**
It is particularly important in a case control study that the exposure time was sufficient enough to show an association between the exposure and the outcome. It may be that the exposure period may be too short or too long to influence the outcome.

10 **Was appropriate statistical analysis used?**
As with any consideration of statistical analysis, consideration should be given to whether there was a more appropriate alternate statistical method that could have been used. The methods section should be detailed enough for reviewers to identify which analytical techniques were used (in particular, regression or stratification) and how specific confounders were measured.

For studies utilising regression analysis, it is useful to identify if the study identified which variables were includ3ed and how they related to the outcome. If stratification was the analytical approach used, were the strata of analysis defined by the specified variables? Additionally, it is also important to assess the appropriateness of the analytical strategy in terms of the assumptions associated with the approach as differing methods of analysis are based on differing assumptions about the data and how it will respond.

Checklist for case reports

Introduction

JBI is an international research organisation based in the Faculty of Health and Medical Sciences at the University of Adelaide, South Australia. JBI develops and delivers unique evidence-based information, software, education and training designed to improve healthcare practice and health outcomes. With over 70 Collaborating Entities, servicing over 90 countries, JBI is a recognised global leader in evidence-based healthcare.

JBI systematic reviews

The core of evidence synthesis is the systematic review of literature of a particular intervention, condition or issue. The systematic review is essentially an analysis of the available literature (i.e., evidence) and a judgment of the effectiveness or otherwise of a practice, involving a series of complex steps. JBI takes a particular view on what counts as evidence and the methods utilised to synthesise those different types of evidence. In line with this broader view of evidence, JBI has developed theories, methodologies and rigorous processes for the critical appraisal and synthesis of these diverse forms of evidence in order to aid in clinical decision-making in healthcare. There now exists JBI guidance for conducting reviews of effectiveness research, qualitative research, prevalence/incidence, aetiology/risk, economic evaluations, text/opinion, diagnostic test accuracy, mixed-methods, umbrella reviews and scoping reviews. Further information regarding JBI systematic reviews can be found in the JBI Evidence Synthesis Manual.

JBI critical appraisal tools

All systematic reviews incorporate a process of critique or appraisal of the research evidence. The purpose of this appraisal is to assess the methodological quality of a study and to determine the extent to which a study has addressed the possibility of bias in its design, conduct and analysis. All papers selected for inclusion in the systematic review (i.e., those that meet the inclusion criteria described in the protocol) need to be subjected to rigorous appraisal by two critical appraisers. The results of this appraisal can then be used to inform synthesis and interpretation of the results of the study. JBI critical appraisal tools have been developed by the JBI and collaborators and approved by the JBI Scientific Committee following extensive peer review. Although designed for use in systematic reviews, JBI critical appraisal tools can also be used when creating Critically Appraised Topics (CAT), in journal clubs and as an educational tool.

JBI critical appraisal checklist for case reports

Reviewer_____ Date _____

Author _____ Year _____ Record Number _____

	Yes	No	Unclear	Not applicable
1 Were patient's demographic characteristics clearly described?	□	□	□	□
2 Was the patient's history clearly described and presented as a timeline?	□	□	□	□
3 Was the current clinical condition of the patient on presentation clearly described?	□	□	□	□
4 Were diagnostic tests or assessment methods and the results clearly described?	□	□	□	□
5 Was the intervention(s) or treatment procedure(s) clearly described?	□	□	□	□
6 Was the post-intervention clinical condition clearly described?	□	□	□	□
7 Were adverse events (harms) or unanticipated events identified and described?	□	□	□	□
8 Does the case report provide takeaway lessons?	□	□	□	□

Overall appraisal: Include □ Exclude □ Seek further info □
Comments (Including reason for exclusion)

_____

_____

_____

Explanation of case reports critical appraisal

How to cite: Moola S, Munn Z, Tufanaru C, Aromataris E, Sears K, Sfetcu R, Currie M, Qureshi R, Mattis P, Lisy K, Mu P-F. Chapter 7: Systematic reviews of etiology and risk. In: Aromataris E, Munn Z (Editors). JBI Manual for Evidence Synthesis. JBI, 2020. Available from https://synthesismanual.jbi.global

Case reports critical appraisal tool

Answers: Yes, No, Unclear or Not/Applicable

1 **Were patient's demographic characteristics clearly described?**
Does the case report clearly describe patient's age, sex, race, medical history, diagnosis, prognosis, previous treatments, past and current diagnostic test results, and medications? The setting and context may also be described.

2   Was the patient's history clearly described and presented as a timeline?
A good case report will clearly describe the history of the patient, their medical, family and psychosocial history including relevant genetic information, as well as relevant past interventions and their outcomes. (CARE Checklist 2013)

3   Was the current clinical condition of the patient on presentation clearly described?
The current clinical condition of the patient should be described in detail including the uniqueness of the condition/disease, symptoms, frequency and severity. The case report should also be able to present whether differential diagnoses was considered.

4   Were diagnostic tests or methods and the results clearly described?
A reader of the case report should be provided sufficient information to understand how the patient was assessed. It is important that all appropriate tests are ordered to confirm a diagnosis and therefore the case report should provide a clear description of various diagnostic tests used (whether a gold standard or alternative diagnostic tests). Photographs or illustrations of diagnostic procedures, radiographs, or treatment procedures are usually presented when appropriate to convey a clear message to readers.

5   Was the intervention(s) or treatment procedure(s) clearly described?
It is important to clearly describe treatment or intervention procedures as other clinicians will be reading the paper and therefore may enable clear understanding of the treatment protocol. The report should describe the treatment/intervention protocol in detail; for example, in pharmacological management of dental anxiety – the type of drug, route of administration, drug dosage and frequency, and any side effects.

6   Was the post-intervention clinical condition clearly described?
A good case report should clearly describe the clinical condition post-intervention in terms of the presence or lack thereof symptoms. The outcomes of management/treatment when presented as images or figures would help in conveying the information to the reader/clinician.

7   Were adverse events (harms) or unanticipated events identified and described?
With any treatment/intervention/drug, there are bound to be some adverse events and in some cases, they may be severe. It is important that adverse events are clearly documented and described, particularly when a new or unique condition is being treated or when a new drug or treatment is used. In addition, unanticipated events, if any that may yield new or useful information should be identified and clearly described.

8   Does the case report provide takeaway lessons?
Case reports should summarise key lessons learned from a case in terms of the background of the condition/disease and clinical practice guidance for clinicians when presented with similar cases.

Reference

Gagnier, J. J., Kienle, G., Altman, D. G., Moher, D., Sox, H., Riley, D., CARE Group. (2013). The CARE guidelines: Consensus-based clinical case reporting guideline development. *Headache: The Journal of Head and Face Pain, 53*(10), 1541–1547.

Checklist for case series

Introduction

JBI is an international research organisation based in the Faculty of Health and Medical Sciences at the University of Adelaide, South Australia. JBI develops and delivers unique evidence-based information, software, education and training designed to improve healthcare practice and health outcomes. With over 70 Collaborating Entities, servicing over 90 countries, JBI is a recognised global leader in evidence-based healthcare.

JBI systematic reviews

The core of evidence synthesis is the systematic review of literature of a particular intervention, condition or issue. The systematic review is essentially an analysis of the available literature (i.e., evidence) and a judgment of the effectiveness or otherwise of a practice, involving a series of complex steps. JBI takes a particular view on what counts as evidence and the methods utilised to synthesise those different types of evidence. In line with this broader view of evidence, JBI has developed theories, methodologies and rigorous processes for the critical appraisal and synthesis of these diverse forms of evidence in order to aid in clinical decision-making in healthcare. There now exists JBI guidance for conducting reviews of effectiveness research, qualitative research, prevalence/incidence, aetiology/risk, economic evaluations, text/opinion, diagnostic test accuracy, mixed-methods, umbrella reviews and scoping reviews. Further information regarding JBI systematic reviews can be found in the JBI Evidence Synthesis Manual.

JBI critical appraisal tools

All systematic reviews incorporate a process of critique or appraisal of the research evidence. The purpose of this appraisal is to assess the methodological quality of a study and to determine the extent to which a study has addressed the possibility of bias in its design, conduct and analysis. All papers selected for inclusion in the systematic review (i.e., those that meet the inclusion criteria described in the protocol) need to be subjected to rigorous appraisal by two critical appraisers. The results of this appraisal can then be used to inform synthesis and interpretation of the results of the study. JBI critical appraisal tools have been developed by the JBI and collaborators and approved by the JBI Scientific Committee following extensive peer review. Although designed for use in systematic reviews, JBI critical appraisal tools can also be used when creating Critically Appraised Topics (CAT), in journal clubs and as an educational tool.

JBI critical appraisal checklist for case series

Reviewer _____ Date _____

Author _____ Year _____ Record Number _____

	Yes	No	Unclear	Not applicable
1 Were there clear criteria for inclusion in the case series?	□	□	□	□
2 Was the condition measured in a standard, reliable way for all participants included in the case series?	□	□	□	□
3 Were valid methods used for identification of the condition for all participants included in the case series?	□	□	□	□
4 Did the case series have consecutive inclusion of participants?	□	□	□	□
5 Did the case series have complete inclusion of participants?	□	□	□	□
6 Was there clear reporting of the demographics of the participants in the study?	□	□	□	□
7 Was there clear reporting of clinical information of the participants?	□	□	□	□
8 Were the outcomes or follow up results of cases clearly reported?	□	□	□	□
9 Was there clear reporting of the presenting site(s)/clinic(s) demographic information?	□	□	□	□
10 Was statistical analysis appropriate?	□	□	□	□

Overall appraisal: Include □   Exclude □   Seek further info □
Comments (Including reason for exclusion)

_____

_____

_____

Introduction to the case series critical appraisal tool

How to cite: Munn Z, Barker T, Moola S, Tufanaru C, Stern C, McArthur A, Stephenson M, Aromataris E. Methodological quality of case series studies, JBI Evidence Synthesis, doi: 10.11124/JBISRIR-D-19-00099

The definition of a case series varies across the medical literature, which has resulted in inconsistent use of this term (Appendix 3.1).[1-3] The gamut of case studies is wide, with some studies claiming to be a case series realistically being nothing more than a collection of case reports, with others more akin to cohort studies or even quasi-experimental before and after studies. This has created difficulty in assigning 'case series' a position in the hierarchy of evidence and identifying and appropriate critical appraisal tool.[1,2]

Dekkers et al. define a case series as a study in which 'only patients with the outcome are sampled (either those who have an exposure or those who are selected without regard to exposure), which does not permit calculation of an absolute risk'.[1p.39] The outcome could be a disease or a disease related outcome. This is contrasted to cohort studies where sampling is based on exposure (or characteristic), and case control studies where there is a comparison group without the disease.

The completeness of a case series contributes to its reliability.[1] Studies that indicate a consecutive and complete inclusion are more reliable than those that do not. For example, a case series that states 'we included all patients (24) with osteosarcoma who presented to our clinic between March 2005 and June 2006' is more reliable than a study that simply states 'we report a case series of 24 people with osteosarcoma'.

For the purposes of this checklist, we agree with the principles outlined in the Dekker et al. paper, and define case series as studies where only patients with a certain disease or disease-related outcome are sampled. Some of the items below relate to risk of bias, whilst others relate to ensuring adequate reporting and statistical analysis. A response of 'no' to any of the questions below negatively impacts the quality of a case series.

Tool guidance

Answers: Yes, No, Unclear or Not/Applicable

1 **Were there clear criteria for inclusion in the case series?**
   The authors should provide clear inclusion (and exclusion criteria where appropriate) for the study participants. The inclusion/exclusion criteria should be specified (e.g., risk and stage of disease progression) with sufficient detail and all the necessary information critical to the study.

2 **Was the condition measured in a standard, reliable way for all participants included in the case series?**
   The study should clearly describe the method of measurement of the condition. This should be done in a standard (i.e. same way for all patients) and reliable (i.e. repeatable and reproducible results) way.

3 **Were valid methods used for identification of the condition for all participants included in the case series?**
   Many health problems are not easily diagnosed or defined and some measures may not be capable of including or excluding appropriate levels or stages of the health problem. If the outcomes were assessed based on existing definitions or diagnostic criteria, then the answer to this question is likely to be yes. If the outcomes were assessed using observer reported, or self-reported scales, the risk of over- or under-reporting is increased, and objectivity is compromised. Importantly, determine if the measurement tools used were validated instruments as this has a significant impact on outcome assessment validity.

4 **Did the case series have consecutive inclusion of participants?**
   Studies that indicate a consecutive inclusion are more reliable than those that do not. For example, a case series that states 'we included all patients (24) with osteosarcoma who presented to our clinic between March 2005 and June 2006' is more reliable than a study that simply states 'we report a case series of 24 people with osteosarcoma'.

5 **Did the case series have complete inclusion of participants?**
   The completeness of a case series contributes to its reliability (1). Studies that indicate a complete inclusion are more reliable than those that do not. A stated above, a case series that

states 'we included all patients (24) with osteosarcoma who presented to our clinic between March 2005 and June 2006' is more reliable than a study that simply states 'we report a case series of 24 people with osteosarcoma'.

6   Was there clear reporting of the demographics of the participants in the study?

The case series should clearly describe relevant participant's demographics such as the following information where relevant: participant's age, sex, education, geographic region, ethnicity, time period, education.

7   Was there clear reporting of clinical information of the participants?

There should be clear reporting of clinical information of the participants such as the following information where relevant: disease status, comorbidities, stage of disease, previous interventions/treatment, results of diagnostic tests, etc.

8   Were the outcomes or follow-up results of cases clearly reported?

The results of any intervention or treatment should be clearly reported in the case series. A good case study should clearly describe the clinical condition post-intervention in terms of the presence or lack of symptoms. The outcomes of management/treatment when presented as images or figures can help in conveying the information to the reader/clinician. It is important that adverse events are clearly documented and described, particularly a new or unique condition is being treated or when a new drug or treatment is used. In addition, unanticipated events, if any that may yield new or useful information should be identified and clearly described.

9   Was there clear reporting of the presenting site(s)/clinic(s) demographic information?

Certain diseases or conditions vary in prevalence across different geographic regions and populations (e.g. women vs. men, sociodemographic variables between countries). The study sample should be described in sufficient detail so that other researchers can determine if it is comparable to the population of interest to them.

10   Was statistical analysis appropriate?

As with any consideration of statistical analysis, consideration should be given to whether there was a more appropriate alternate statistical method that could have been used. The methods section of studies should be detailed enough for reviewers to identify which analytical techniques were used and whether these were suitable.

References

1 Dekkers, O. M., Egger, M., Altman, D.G., & Vandenbroucke, J. P. (2012). Distinguishing case series from cohort studies. *Annals of Internal Medicine, 156*(1 Part 1), 37–40.

2 Esene, I. N., Ngu, J., El Zoghby, M., Solaroglu, I., Sikod, A. M., Kotb, A. et al. (2014). Case series and descriptive cohort studies in neurosurgery: The confusion and solution. *Child's Nervous System, 30*(8), 1321–1332.

3 Abu-Zidan, F. M., Abbas, A. K., Hefny, A. F. (2012). Clinical "case series": A concept analysis. *African Health Sciences, 12*(4), 557–562.

4 Straus, S. E., Richardson, W. S., Glasziou, P., & Haynes, R. B. (2005). *Evidence-Based Medicine: How to Practice and Teach EBM* (3rd ed.). Elsevier.

Appendix 3.1 Case series definitions

'A report on a series of patients with an outcome of interest. No control group is involved'.(4) [p.279]

'A case series is a descriptive study involving a group of patients who all have the same disease or condition: the aim is to describe common and differing characteristics of a particular group of individuals' (Oxford Handbook of medical statistics)

'A group or series of case reports involving patients who were given similar treatment. Reports of case series usually contain detailed information about the individual patients. This includes demographic information (e.g., age, gender, and ethnic origin) and information on diagnosis, treatment, response to treatment, and follow-up after treatment'. Law K, Howick J. OCEBM Table of Evidence Glossary. 2013 [cited 2014 10th January]; Available from: http://www.cebm.net/index.aspx?o=1116

'A **case series** (also known as a clinical **series**) is a type of medical research study that tracks subjects with a known exposure, such as patients who have received a similar treatment, or examines their medical records for exposure and outcome'. Wikipedia

'A study which makes observations on a series of individuals, usually all receiving the same intervention, with no control group. Comments: At this stage it is unclear whether case series should be included in Cochrane systematic reviews, but we have left them in the list so that working groups can consider whether there are circumstances in which it would be appropriate to include them, and to assess risk of bias. A particular reason for including case series might be where they provide evidence relating to adverse effects of an intervention. Potential examples of risk of bias might be that if a case series does not [attempt to] recruit consecutive participants, this might introduce a risk of selection bias, while some case series could be at risk of detection bias, if the circumstances in which adverse effects are reported (or elicited) are not standardised'. http://bmg.cochrane.org/research-projectscochrane-risk-bias-tool

Checklist for cohort studies

Introduction

JBI is an international research organisation based in the Faculty of Health and Medical Sciences at the University of Adelaide, South Australia. JBI develops and delivers unique evidence-based information, software, education and training designed to improve healthcare practice and health outcomes. With over 70 Collaborating Entities, servicing over 90 countries, JBI is a recognised global leader in evidence-based healthcare.

JBI systematic reviews

The core of evidence synthesis is the systematic review of literature of a particular intervention, condition or issue. The systematic review is essentially an analysis of the available literature (i.e., evidence) and a judgment of the effectiveness or otherwise of a practice, involving a series of complex steps. JBI takes a particular view on what counts as evidence and the methods utilised to synthesise those different types of evidence. In line with this broader view of evidence, JBI has developed theories, methodologies and rigorous processes for the critical appraisal and synthesis of these diverse forms of evidence in order to aid in clinical decision-making in healthcare. There now exists JBI guidance for conducting reviews of effectiveness research, qualitative research, prevalence/incidence, aetiology/risk, economic evaluations, text/opinion, diagnostic test accuracy, mixed-methods, umbrella reviews and scoping reviews. Further information regarding JBI systematic reviews can be found in the JBI Evidence Synthesis Manual.

JBI critical appraisal tools

All systematic reviews incorporate a process of critique or appraisal of the research evidence. The purpose of this appraisal is to assess the methodological quality of a study and to determine the extent to which a study has addressed the possibility of bias in its design, conduct and analysis. All papers selected for inclusion in the systematic review (i.e., those that meet the inclusion criteria described in the protocol) need to be subjected to rigorous appraisal by two critical appraisers. The results of this appraisal can then be used to inform synthesis and interpretation of the results of the study. JBI critical appraisal tools have been developed by the JBI and collaborators and approved by the JBI Scientific Committee following extensive peer review. Although designed for use in systematic reviews, JBI critical appraisal tools can also be used when creating Critically Appraised Topics (CAT), in journal clubs and as an educational tool.

JBI critical appraisal checklist for cohort studies

Reviewer_____ Date_____

Author_____ Year_____ Record Number_____

	Yes	No	Unclear	Not applicable
1 Were the two groups similar and recruited from the same population?	☐	☐	☐	☐
2 Were the exposures measured similarly to assign people to both exposed and unexposed groups?	☐	☐	☐	☐
3 Was the exposure measured in a valid and reliable way?	☐	☐	☐	☐
4 Were confounding factors identified?	☐	☐	☐	☐
5 Were strategies to deal with confounding factors stated?	☐	☐	☐	☐
6 Were the groups/participants free of the outcome at the start of the study (or at the moment of exposure)?	☐	☐	☐	☐
7 Were the outcomes measured in a valid and reliable way?	☐	☐	☐	☐
8 Was the follow up time reported and sufficient to be long enough for outcomes to occur?	☐	☐	☐	☐
9 Was follow up complete, and if not, were the reasons to loss to follow up described and explored?	☐	☐	☐	☐
10 Were strategies to address incomplete follow up utilised?	☐	☐	☐	☐
11 Was appropriate statistical analysis used?	☐	☐	☐	☐

Overall appraisal: Include ☐   Exclude ☐   Seek further info ☐
Comments (Including reason for exclusion)

_____

_____

_____

Explanation of cohort studies critical appraisal

How to Cite: *Moola S, Munn Z, Tufanaru C, Aromataris E, Sears K, Sfetcu R, Currie M, Qureshi R, Mattis P, Lisy K, Mu P-F. Chapter 7: Systematic reviews of etiology and risk. In: Aromataris E, Munn Z (Editors). JBI Manual for Evidence Synthesis.* JBI, 2020. Available from https://synthesismanual.jbi.global

Cohort studies critical appraisal tool

Answers: Yes, No, Unclear or Not/Applicable

1 Were the two groups similar and recruited from the same population?
Check the paper carefully for descriptions of participants to determine if patients within and across groups have similar characteristics in relation to exposure (e.g., risk factor under investigation). The two groups selected for comparison should be as similar as possible in all characteristics except for their exposure status, relevant to the study in question. The authors should provide clear inclusion and exclusion criteria that they developed prior to recruitment of the study participants.

2 Were the exposures measured similarly to assign people to both exposed and unexposed groups?
A high-quality study at the level of cohort design should mention or describe how the exposures were measured. The exposure measures should be clearly defined and described in detail. This will enable reviewers to assess whether or not the participants received the exposure of interest.

3 Was the exposure measured in a valid and reliable way?
The study should clearly describe the method of measurement of exposure. Assessing validity requires that a 'gold standard' is available to which the measure can be compared. The validity of exposure measurement usually relates to whether a current measure is appropriate or whether a measure of past exposure is needed.

Reliability refers to the processes included in an epidemiological study to check repeatability of measurements of the exposures. These usually include intra-observer reliability and inter-observer reliability.

4 Were confounding factors identified?
Confounding has occurred where the estimated intervention exposure effect is biased by the presence of some difference between the comparison groups (apart from the exposure investigated/of interest). Typical confounders include baseline characteristics, prognostic factors, or concomitant exposures (e.g. smoking). A confounder is a difference between the comparison groups and it influences the direction of the study results. A high-quality study at the level of cohort design will identify the potential confounders and measure them (where possible). This is difficult for studies where behavioural, attitudinal or lifestyle factors may impact on the results.

5 Were strategies to deal with confounding factors stated?
Strategies to deal with effects of confounding factors may be dealt within the study design or in data analysis. By matching or stratifying sampling of participants, effects of confounding factors can be adjusted for. When dealing with adjustment in data analysis, assess the statistics used in the study. Most will be some form of multivariate regression analysis to account for the confounding factors measured. Look out for a description of statistical methods as regression methods such as logistic regression are usually employed to deal with confounding factors/variables of interest.

6 Were the groups/participants free of the outcome at the start of the study (or at the moment of exposure)?
The participants should be free of the outcomes of interest at the start of the study. Refer to the 'methods' section in the paper for this information, which is usually found in descriptions of participant/sample recruitment, definitions of variables, and/or inclusion/exclusion criteria.

7 **Were the outcomes measured in a valid and reliable way?**

Read the methods section of the paper. If, for example, lung cancer is assessed based on existing definitions or diagnostic criteria, then the answer to this question is likely to be yes. If lung cancer is assessed using observer reported, or self-reported scales, the risk of over- or under-reporting is increased, and objectivity is compromised. Importantly, determine if the measurement tools used were validated instruments as this has a significant impact on outcome assessment validity.

Having established the objectivity of the outcome measurement (e.g. lung cancer) instrument, it's important to establish how the measurement was conducted. Were those involved in collecting data trained or educated in the use of the instrument/s? (e.g. radiographers). If there was more than one data collector, were they similar in terms of level of education, clinical or research experience, or level of responsibility in the piece of research being appraised?

8 **Was the follow up time reported and sufficient to be long enough for outcomes to occur?**

The appropriate length of time for follow up will vary with the nature and characteristics of the population of interest and/or the intervention, disease or exposure. To estimate an appropriate duration of follow up, read across multiple papers and take note of the range for duration of follow up. The opinions of experts in clinical practice or clinical research may also assist in determining an appropriate duration of follow up. For example, a longer timeframe may be needed to examine the association between occupational exposure to asbestos and the risk of lung cancer. It is important, particularly in cohort studies that follow up is long enough to enable the outcomes. However, it should be remembered that the research question and outcomes being examined would probably dictate the follow up time.

9 **Was follow up complete, and if not, were the reasons to loss to follow up described and explored?**

It is important in a cohort study that a greater percentage of people are followed up. As a general guideline, at least 80% of patients should be followed up. Generally a dropout rate of 5% or less is considered insignificant. A rate of 20% or greater is considered to significantly impact on the validity of the study. However, in observational studies conducted over a lengthy period of time a higher dropout rate is to be expected. A decision on whether to include or exclude a study because of a high dropout rate is a matter of judgement based on the reasons why people dropped out, and whether dropout rates were comparable in the exposed and unexposed groups.

Reporting of efforts to follow up participants that dropped out may be regarded as an indicator of a well conducted study. Look for clear and justifiable description of why people were left out, excluded, dropped out, etc. If there is no clear description or a statement in this regards, this will be a 'No'.

10 **Were strategies to address incomplete follow up utilised?**

Some people may withdraw due to change in employment or some may die; however, it is important that their outcomes are assessed. Selection bias may occur as a result of incomplete follow up. Therefore, participants with unequal follow up periods must be taken into account in the analysis, which should be adjusted to allow for differences in length of follow up periods. This is usually done by calculating rates which use person-years at risk, that is, considering time in the denominator.

11 Was appropriate statistical analysis used?

As with any consideration of statistical analysis, consideration should be given to whether there was a more appropriate alternate statistical method that could have been used. The methods section of cohort studies should be detailed enough for reviewers to identify which analytical techniques were used (in particular, regression or stratification) and how specific confounders were measured.

For studies utilising regression analysis, it is useful to identify if the study identified which variables were included and how they related to the outcome. If stratification was the analytical approach used, were the strata of analysis defined by the specified variables? Additionally, it is also important to assess the appropriateness of the analytical strategy in terms of the assumptions associated with the approach as differing methods of analysis are based on differing assumptions about the data and how it will respond.

Checklist for diagnostic test accuracy studies

Introduction

JBI is an international research organisation based in the Faculty of Health and Medical Sciences at the University of Adelaide, South Australia. JBI develops and delivers unique evidence-based information, software, education and training designed to improve healthcare practice and health outcomes. With over 70 Collaborating Entities, servicing over 90 countries, JBI is a recognised global leader in evidence-based healthcare.

JBI systematic reviews

The core of evidence synthesis is the systematic review of literature of a particular intervention, condition or issue. The systematic review is essentially an analysis of the available literature (i.e., evidence) and a judgment of the effectiveness or otherwise of a practice, involving a series of complex steps. JBI takes a particular view on what counts as evidence and the methods utilised to synthesise those different types of evidence. In line with this broader view of evidence, JBI has developed theories, methodologies and rigorous processes for the critical appraisal and synthesis of these diverse forms of evidence in order to aid in clinical decision-making in healthcare. There now exists JBI guidance for conducting reviews of effectiveness research, qualitative research, prevalence/incidence, aetiology/risk, economic evaluations, text/opinion, diagnostic test accuracy, mixed-methods, umbrella reviews and scoping reviews. Further information regarding JBI systematic reviews can be found in the JBI Evidence Synthesis Manual.

JBI critical appraisal tools

All systematic reviews incorporate a process of critique or appraisal of the research evidence. The purpose of this appraisal is to assess the methodological quality of a study and to determine the extent to which a study has addressed the possibility of bias in its design, conduct and analysis. All papers selected for inclusion in the systematic review (i.e., those that meet the inclusion criteria described in the protocol) need to be subjected to rigorous appraisal by two critical appraisers. The results of this appraisal can then be used to inform synthesis and interpretation of the results of the study. JBI critical appraisal tools have been developed by the JBI and collaborators and approved by the JBI Scientific Committee following extensive peer review. Although designed for use in systematic reviews, JBI critical appraisal tools can also be used when creating Critically Appraised Topics (CAT), in journal clubs and as an educational tool.

JBI critical appraisal checklist for diagnostic test accuracy studies

Reviewer_____ Date_____

Author_____ Year_____ Record Number_____

	Yes	No	Unclear	Not applicable
1 Was a consecutive or random sample of patients enrolled?	☐	☐	☐	☐
2 Was a case control design avoided?	☐	☐	☐	☐
3 Did the study avoid inappropriate exclusions?	☐	☐	☐	☐
4 Were the index test results interpreted without knowledge of the results of the reference standard?	☐	☐	☐	☐
5 If a threshold was used, was it pre-specified?	☐	☐	☐	☐
6 Is the reference standard likely to correctly classify the target condition?	☐	☐	☐	☐
7 Were the reference standard results interpreted without knowledge of the results of the index test?	☐	☐	☐	☐
8 Was there an appropriate interval between index test and reference standard?	☐	☐	☐	☐
9 Did all patients receive the same reference standard?	☐	☐	☐	☐
10 Were all patients included in the analysis?	☐	☐	☐	☐

Overall appraisal: Include ☐   Exclude ☐   Seek further info ☐
Comments (Including reason for exclusion)

_____
_____
_____

Diagnostic test accuracy studies critical appraisal tool

How to cite: *Whiting PF, Rutjes AW, Westwood ME, Mallett S, Deeks JJ, Reitsma JB, Leeflang MM, Sterne JA, Bossuyt PM, QUADAS-2 Group. QUADAS-2: a revised tool for the quality assessment of diagnostic accuracy studies. Ann Intern Med. 2011;155(8):529–36.*

Campbell JM, Klugar M, Ding S, Carmody DP, Hakonsen SJ, Jadotte YT, White S, Munn Z. Diagnostic test accuracy: methods for systematic review and meta-analysis. Int J Evid Based Healthc. 2015;13(3):154–62.

Answers: Yes, No, Unclear or Not/Applicable

Patient selection

1 Was a consecutive or random sample of patients enrolled?
Studies should state or describe their method of enrolment. If it is claimed that a random sample was chosen the method of randomisation should be stated (and appropriate). It is acceptable if studies do not say 'consecutive' but instead describe consecutive enrolment; that is, 'all patients from …. till …. were included'.

2 Was a case control design avoided?
Case control studies are described in detail in the reviewers manual. In essence, if a study design involves recruiting participants who are already known by other means to have the

diagnosis of interest and investigating whether the test of interest correctly identifies them as such, the answer is 'No'.

3 Did the study avoid inappropriate exclusions?

If patients are excluded for reasons that would likely influence the conduct, interpretation or results of the test, this may bias the results. Examples include: excluding patients on which the test is difficult to conduct, excluding patients with borderline results, excluding patients with clear clinical indicators of the diagnosis of interest.

Index test

4 Were the index test results interpreted without knowledge of the results of the reference standard?

The results of the index test should be interpreted by someone who is blind to the results of the reference test. The reference test may not have been conducted at the point that the index test is carried out, if so the answer to this question will be 'Yes'. If the person who interprets the index test also interpreted the reference test then it is assumed that this question will be answered 'No' unless there are other factors in play (for instance, the interpretation of the results may be separate from their collection, in which case the interpreter may be blinded to patient identity and past reference test results).

5 If a threshold was used, was it pre-specified?

Diagnostic thresholds may be chosen based on what gives the optimum accuracy from the data, or they may be pre-specified. When no diagnostic threshold is applied (i.e. the results of a test is based on the observation of a specific characteristic which is either there or not) this question will be answered NA.

6 Is the reference standard likely to correctly classify the target condition?

The reference test should be the gold standard for the diagnosis of the condition of interest. Additionally, the reporting of the study should describe its conduct in sufficient detail that the reviewers can be confident that it has been correctly and competently implemented.

7 Were the reference standard results interpreted without knowledge of the results of the index test?

The points made for criteria 4 apply equally here. The results of the reference test should be interpreted by someone who is blind to the results of the index test. The index test may not have been conducted at the point that the reference test is carried out, if so the answer to this question will be 'Yes'. If the person who interprets the reference test also interpreted the index test, then it is assumed that this question will be answered 'No' unless there are other factors in play (for instance, the interpretation of the results may be separate from their collection, in which case the interpreter may be blinded to patient identity and past index test results).

8 Was there an appropriate interval between index test and reference standard?

The index test and the reference test should be carried out close enough together that the status of the patient could not have meaningfully changed. The maximum acceptable time will vary based on characteristics of the population and condition of interest.

9 **Did all patients receive the same reference standard?**
The reference standard by which patients are classed as having or not having the condition of interest should be the same for all patients. If the results of the index test influence how or whether the reference test is used (i.e. where an apparent false negative may be detected the study design may call for a 'double check') this may result in biased estimates of sensitivity and specificity. Additionally, in some studies two parallel reference tests may be used (on different patients) and the results then pooled. In either case the results should be 'No'.

10 **Were all patients included in the analysis?**
Loses to follow up should be explained and there cause and frequency should be considered in whether they are likely to have had an effect on the results (Subjectivity may exist in this context, overall low tolerance should be applied in deciding to answer 'No' to this question, but a single withdrawal from a large cohort should not necessarily force a negative response). However, if a patient's results being difficult to interpret causes their data to be excluded from the analysis this will exaggerate the estimate of DTA, and this question should definitely be answered 'No'.

Checklist for economic evaluations

Introduction

JBI is an international research organisation based in the Faculty of Health and Medical Sciences at the University of Adelaide, South Australia. JBI develops and delivers unique evidence-based information, software, education and training designed to improve healthcare practice and health outcomes. With over 70 Collaborating Entities, servicing over 90 countries, JBI is a recognised global leader in evidence-based healthcare.

JBI systematic reviews

The core of evidence synthesis is the systematic review of literature of a particular intervention, condition or issue. The systematic review is essentially an analysis of the available literature (i.e., evidence) and a judgment of the effectiveness or otherwise of a practice, involving a series of complex steps. JBI takes a particular view on what counts as evidence and the methods utilised to synthesise those different types of evidence. In line with this broader view of evidence, JBI has developed theories, methodologies and rigorous processes for the critical appraisal and synthesis of these diverse forms of evidence in order to aid in clinical decision-making in healthcare. There now exists JBI guidance for conducting reviews of effectiveness research, qualitative research, prevalence/incidence, aetiology/risk, economic evaluations, text/opinion, diagnostic test accuracy, mixed-methods, umbrella reviews and scoping reviews. Further information regarding JBI systematic reviews can be found in the JBI Evidence Synthesis Manual.

JBI critical appraisal tools

All systematic reviews incorporate a process of critique or appraisal of the research evidence. The purpose of this appraisal is to assess the methodological quality of a study and to determine the extent to which a study has addressed the possibility of bias in its design, conduct and analysis. All papers selected for inclusion in the systematic review (i.e., those that meet the inclusion criteria described in the protocol) need to be subjected to rigorous appraisal by two critical appraisers. The results of this appraisal can then be used to inform synthesis and interpretation of the results of the study. JBI critical appraisal tools have been developed by the JBI and collaborators and approved by the JBI Scientific Committee following extensive peer review. Although designed for use in systematic reviews, JBI critical appraisal tools can also be used when creating Critically Appraised Topics (CAT), in journal clubs and as an educational tool.

JBI critical appraisal checklist for economic evaluations

Reviewer _____ Date _____

Author _____ Year _____ Record Number _____

	Yes	No	Unclear	Not applicable
1 Is there a well-defined question?	□	□	□	□
2 Is there comprehensive description of alternatives?	□	□	□	□
3 Are all important and relevant costs and outcomes for each alternative identified?	□	□	□	□
4 Has clinical effectiveness been established?	□	□	□	□
5 Are costs and outcomes measured accurately?	□	□	□	□
6 Are costs and outcomes valued credibly?	□	□	□	□
7 Are costs and outcomes adjusted for differential timing?	□	□	□	□
8 Is there an incremental analysis of costs and consequences?	□	□	□	□
9 Were sensitivity analyses conducted to investigate uncertainty in estimates of cost or consequences?	□	□	□	□
10 Do study results include all issues of concern to users?	□	□	□	□
11 Are the results generalisable to the setting of interest in the review?	□	□	□	□

Overall appraisal: Include □   Exclude □   Seek further info □

Comments (Including reason for exclusion)

_____

_____

_____

JBI critical appraisal checklist for economic evaluations

How to cite: Gomersall JS, Jadotte YT, Xue Y, Lockwood S, Riddle D, Preda A. Conducting systematic reviews of economic evaluations. Int J Evid Based Healthc. 2015;13(3):170–178.

This tool is informed by the work of Drummond et al, Methods for the economic evaluation of health care programmes. 2nd Edition. Oxford: Oxford Medical Publications, 1997.

1 Is there a well-defined question/objective?

Consider the following before marking the study as compliant with this quality criterion:

- Is the objective/question of the study clearly stated?
- Does the statement reflect the perspective (e.g. patient or community or societal or health provider) used in measurement of costs or/and cost effectiveness?
- Was the study placed in a particular decision-making context?

2 Is there a comprehensive description of alternatives?

To be marked as compliant with this criterion the authors of the study should offer a clear description of the intervention or interventions considered in the economic evaluation and the comparator or comparators. Compliance does not require that a broad range of interventions and comparators was considered. What is important here is clear description of the nature of the intervention and comparator whose cost/effeteness was measured.

3 Are all important and relevant costs and outcomes for each alternative identified?

This quality criterion assesses the comprehensiveness and relevant of the cost and cost effectiveness outcomes measured in the economic evaluation. When deciding whether all important costs and outcomes have been identified/measured in the study reflect on whether the outcomes are sufficient in light of the objectives of the study. It is appropriate for a study that has the objective of measuring a narrow range of costs and benefits to identify and measure a limited range. However, the limits of the narrow approach should be drawn out in the study. It is not appropriate for a study which implies in its objective statement that it measures a broad range of costs for a broad range out outcomes to include only a very limited range of relevant costs and outcomes.

4 Has clinical effectiveness been established?

To assess compliance with this quality criterion requires considering whether the study has reported the evidence used to derive the effectiveness estimate and the level of this evidence. If it is not clear how the effectiveness estimate was derived, the study cannot be marked as compliant. To achieve compliance for this criterion the effectiveness estimate in the evaluation does not need to be derived from the same study as the resource use/cost estimate. What is important is the there is a solid evidence base under-pinning the assumptions about the direction and magnitude of the effectiveness measure(s) used in the evaluation.

5 Are costs and outcomes measured accurately?

This quality criterion assesses whether the study has used appropriate/best practice measurement method to measure costs and effectiveness. To decide whether a study should be marked as compliant consider whether the methods section of the paper offers a detail description of the measures used for costs and outcomes and how it justifies them. In addition, consider whether the authors/study implementers discussed any limitations associated with the measures used and concerns about the accuracy of measurement. In economic evaluations it is often difficult to measure costs and outcomes accurately, and hence in many cases this quality criterion will be difficult to achieve.

6 Are costs and outcomes valued credibly?

This quality criterion assesses whether appropriate prices were used to value costs and the validity of the valuation of benefits. It requires considering the method description and judging where there is a sufficient explanation about how costs and outcomes were valued and whether the justification for it is persuasive.

7 Are costs and outcomes adjusted for differential timing?

To be marked compliant for this question the study should have identified and justified the discount rate used. The time frame over which the study was conducted should also have been identified and justified.

8 Is there any incremental analysis of costs and consequences?

To achieve compliance the paper should report a measure that shows the change in costs and benefits for the intervention and comparator for a marginal shift in resources from the comparator to the intervention.

9 **Were sensitivity analysis conducted to investigate uncertainty in estimates of costs or outcomes?**

Sensitivity analysis is critical for establishing the validity of any economic evaluations results. To be compliant a study must present sensitivity testing results that describe how the study findings vary with changes in key variables (e.g., relative prices and intervention estimates?) conducted to check the robustness of findings.

10 **Do study results include all issues of concern to users?**

This question reflects on the comprehensiveness of coverage in the reporting of results. In deciding whether to mark the study as compliance consider whether the range of measures presented provider answers to all the questions users/decision makers would want to know when taking a decision about whether to implement the programme examined (or cutting it)?

11 **Are the results generalisable to the setting of interest in the review?**

To be marked as compliant for this last quality criterion the paper should: (i) have described the study setting adequately; (ii) discuss the issue of transferability of findings and how the results are generalisable to other settings with similar characteristics

Checklist for prevalence studies

Introduction

JBI is an international research organisation based in the Faculty of Health and Medical Sciences at the University of Adelaide, South Australia. JBI develops and delivers unique evidence-based information, software, education and training designed to improve healthcare practice and health outcomes. With over 70 Collaborating Entities, servicing over 90 countries, JBI is a recognised global leader in evidence-based healthcare.

JBI systematic reviews

The core of evidence synthesis is the systematic review of literature of a particular intervention, condition or issue. The systematic review is essentially an analysis of the available literature (i.e., evidence) and a judgment of the effectiveness or otherwise of a practice, involving a series of complex steps. JBI takes a particular view on what counts as evidence and the methods utilised to synthesise those different types of evidence. In line with this broader view of evidence, JBI has developed theories, methodologies and rigorous processes for the critical appraisal and synthesis of these diverse forms of evidence in order to aid in clinical decision-making in healthcare. There now exists JBI guidance for conducting reviews of effectiveness research, qualitative research, prevalence/incidence, aetiology/risk, economic evaluations, text/opinion, diagnostic test accuracy, mixed-methods, umbrella reviews and scoping reviews. Further information regarding JBI systematic reviews can be found in the JBI Evidence Synthesis Manual.

JBI critical appraisal tools

All systematic reviews incorporate a process of critique or appraisal of the research evidence. The purpose of this appraisal is to assess the methodological quality of a study and to determine the extent to which a study has addressed the possibility of bias in its design, conduct and analysis. All papers selected for inclusion in the systematic review (i.e., those that meet the inclusion criteria described in the protocol) need to be subjected to rigorous appraisal by two critical appraisers. The results of this appraisal can then be used to inform synthesis and interpretation of the results of the study. JBI critical appraisal tools have been developed by the JBI and collaborators and approved by the JBI Scientific Committee following extensive peer review. Although designed for use in systematic reviews, JBI critical appraisal tools can also be used when creating Critically Appraised Topics (CAT), in journal clubs and as an educational tool.

JBI critical appraisal checklist for studies reporting prevalence data

Reviewer _____ Date _____

Author _____ Year _____ Record Number _____

	Yes	No	Unclear	Not applicable
1 Was the sample frame appropriate to address the target population?	☐	☐	☐	☐
2 Were study participants sampled in an appropriate way?	☐	☐	☐	☐
3 Was the sample size adequate?	☐	☐	☐	☐
4 Were the study subjects and the setting described in detail?	☐	☐	☐	☐
5 Was the data analysis conducted with sufficient coverage of the identified sample?	☐	☐	☐	☐
6 Were valid methods used for the identification of the condition?	☐	☐	☐	☐
7 Was the condition measured in a standard, reliable way for all participants?	☐	☐	☐	☐
8 Was there appropriate statistical analysis?	☐	☐	☐	☐
9 Was the response rate adequate, and if not, was the low response rate managed appropriately?	☐	☐	☐	☐

Overall appraisal: Include ☐   Exclude ☐   Seek further info ☐

Comments (Including reason for exclusion)

_____

_____

_____

JBI critical appraisal checklist for studies reporting prevalence data

How to cite: Munn Z, Moola S, Lisy K, Riitano D, Tufanaru C. Methodological guidance for systematic reviews of observational epidemiological studies reporting prevalence and incidence data. Int J Evid Based Healthc. 2015;13(3):147–153.

Answers: Yes, No, Unclear or Not/Applicable

1 Was the sample frame appropriate to address the target population?

This question relies upon knowledge of the broader characteristics of the population of interest and the geographical area. If the study is of women with breast cancer, knowledge of at least the characteristics, demographics and medical history is needed. The term "target population" should not be taken to infer every individual from everywhere or with similar disease or exposure characteristics. Instead, give consideration to specific population characteristics in the study, including age range, gender, morbidities, medications, and other potentially influential factors. For example, a sample frame may not be appropriate to address the target population if a certain group has been used (such as those working for one organisation, or one profession) and the results then inferred to the target population (i.e. working adults). A sample frame may be appropriate when it includes almost all the members of the target population (i.e. a census, or a complete list of participants or complete registry data).

2 Were study participants recruited in an appropriate way?

Studies may report random sampling from a population, and the methods section should report how sampling was performed. Random probabilistic sampling from a defined sub-set of the population (sample frame) should be employed in most cases, however, random probabilistic sampling is not needed when everyone in the sampling frame will be included/ analysed. For example, reporting on all the data from a good census is appropriate as a good census will identify everybody. When using cluster sampling, such as a random sample of villages within a region, the methods need to be clearly stated as the precision of the final prevalence estimate incorporates the clustering effect. Convenience samples, such as a street survey or interviewing lots of people at a public gatherings are not considered to provide a representative sample of the base population.

3 Was the sample size adequate?

The larger the sample, the narrower will be the confidence interval around the prevalence estimate, making the results more precise. An adequate sample size is important to ensure good precision of the final estimate. Ideally we are looking for evidence that the authors conducted a sample size calculation to determine an adequate sample size. This will estimate how many subjects are needed to produce a reliable estimate of the measure(s) of interest. For conditions with a low prevalence, a larger sample size is needed. Also consider sample sizes for subgroup (or characteristics) analyses, and whether these are appropriate. Sometimes, the study will be large enough (as in large national surveys) whereby a sample size calculation is not required. In these cases, sample size can be considered adequate.

When there is no sample size calculation and it is not a large national survey, the reviewers may consider conducting their own sample size analysis using the following formula: (Naing et al., 2006; Daniel, 1999)

$$n = \frac{Z2P(1-P)}{d2}$$

Where:

n = sample size
Z = Z statistic for a level of confidence
P = Expected prevalence or proportion (in proportion of one; if 20%, P = 0.2)
d = precision (in proportion of one; if 5%, d = 0.05)

Ref:

Naing L, Winn T, Rusli BN. Practical issues in calculating the sample size for prevalence studies Archives of Orofacial Sciences. 2006;1:9–14.

Daniel WW. Biostatistics: A Foundation for Analysis in the Health Sciences. Edition. 7th ed. New York: John Wiley & Sons. 1999.

4 Were the study subjects and setting described in detail?

Certain diseases or conditions vary in prevalence across different geographic regions and populations (e.g. Women vs. Men, sociodemographic variables between countries). The study sample should be described in sufficient detail so that other researchers can determine if it is comparable to the population of interest to them.

5   Was data analysis conducted with sufficient coverage of the identified sample?
Coverage bias can occur when not all subgroups of the identified sample respond at the same rate. For instance, you may have a very high response rate overall for your study, but the response rate for a certain subgroup (i.e. older adults) may be quite low.

6   Were valid methods used for the identification of the condition?
Here we are looking for measurement or classification bias. Many health problems are not easily diagnosed or defined and some measures may not be capable of including or excluding appropriate levels or stages of the health problem. If the outcomes were assessed based on existing definitions or diagnostic criteria, then the answer to this question is likely to be yes. If the outcomes were assessed using observer reported, or self-reported scales, the risk of over- or under-reporting is increased, and objectivity is compromised. Importantly, determine if the measurement tools used were validated instruments as this has a significant impact on outcome assessment validity.

7   Was the condition measured in a standard, reliable way for all participants?
Considerable judgment is required to determine the presence of some health outcomes. Having established the validity of the outcome measurement instrument (see item 6 of this scale), it is important to establish how the measurement was conducted. Were those involved in collecting data trained or educated in the use of the instrument/s? If there was more than one data collector, were they similar in terms of level of education, clinical or research experience, or level of responsibility in the piece of research being appraised? When there was more than one observer or collector, was there comparison of results from across the observers? Was the condition measured in the same way for all participants?

8   Was there appropriate statistical analysis?
Importantly, the numerator and denominator should be clearly reported, and percentages should be given with confidence intervals. The methods section should be detailed enough for reviewers to identify the analytical technique used and how specific variables were measured. Additionally, it is also important to assess the appropriateness of the analytical strategy in terms of the assumptions associated with the approach as differing methods of analysis are based on differing assumptions about the data and how it will respond.

9   Was the response rate adequate, and if not, was the low response rate managed appropriately?
A large number of dropouts, refusals or "not founds" amongst selected subjects may diminish a study's validity, as can a low response rates for survey studies. The authors should clearly discuss the response rate and any reasons for non-response and compare persons in the study to those not in the study, particularly with regards to their sociodemographic characteristics. If reasons for non-response appear to be unrelated to the outcome measured and the characteristics of non-responders are comparable to those who do respond in the study (addressed in Question 5, coverage bias), the researchers may be able to justify a more modest response rate.

Checklist for qualitative research

Introduction

JBI is an international research organisation based in the Faculty of Health and Medical Sciences at the University of Adelaide, South Australia. JBI develops and delivers unique evidence-based information, software, education and training designed to improve healthcare practice and health outcomes. With over 70 Collaborating Entities, servicing over 90 countries, JBI is a recognised global leader in evidence-based healthcare.

JBI systematic reviews

The core of evidence synthesis is the systematic review of literature of a particular intervention, condition or issue. The systematic review is essentially an analysis of the available literature (i.e., evidence) and a judgment of the effectiveness or otherwise of a practice, involving a series of complex steps. JBI takes a particular view on what counts as evidence and the methods utilised to synthesise those different types of evidence. In line with this broader view of evidence, JBI has developed theories, methodologies and rigorous processes for the critical appraisal and synthesis of these diverse forms of evidence in order to aid in clinical decision-making in healthcare. There now exists JBI guidance for conducting reviews of effectiveness research, qualitative research, prevalence/incidence, aetiology/risk, economic evaluations, text/opinion, diagnostic test accuracy, mixed-methods, umbrella reviews and scoping reviews. Further information regarding JBI systematic reviews can be found in the JBI Evidence Synthesis Manual.

JBI critical appraisal tools

All systematic reviews incorporate a process of critique or appraisal of the research evidence. The purpose of this appraisal is to assess the methodological quality of a study and to determine the extent to which a study has addressed the possibility of bias in its design, conduct and analysis. All papers selected for inclusion in the systematic review (i.e., those that meet the inclusion criteria described in the protocol) need to be subjected to rigorous appraisal by two critical appraisers. The results of this appraisal can then be used to inform synthesis and interpretation of the results of the study. JBI critical appraisal tools have been developed by the JBI and collaborators and approved by the JBI Scientific Committee following extensive peer review. Although designed for use in systematic reviews, JBI critical appraisal tools can also be used when creating Critically Appraised Topics (CAT), in journal clubs and as an educational tool.

JBI critical appraisal checklist for qualitative research

Reviewer_____ Date _____

Author _____ Year _____ Record Number _____

	Yes	*No*	*Unclear*	*Not applicable*
1 Is there congruity between the stated philosophical perspective and the research methodology?	□	□	□	□
2 Is there congruity between the research methodology and the research question or objectives?	□	□	□	□
3 Is there congruity between the research methodology and the methods used to collect data?	□	□	□	□
4 Is there congruity between the research methodology and the representation and analysis of data?	□	□	□	□
5 Is there congruity between the research methodology and the interpretation of results?	□	□	□	□
6 Is there a statement locating the researcher culturally or theoretically?	□	□	□	□
7 Is the influence of the researcher on the research, and vice versa, addressed?	□	□	□	□
8 Are participants, and their voices, adequately represented?	□	□	□	□
9 Is the research ethical according to current criteria or, for recent studies, and is there evidence of ethical approval by an appropriate body?	□	□	□	□
10 Do the conclusions drawn in the research report flow from the analysis, or interpretation, of the data?	□	□	□	□

Overall appraisal: Include □   Exclude □   Seek further info □
Comments (Including reason for exclusion)

_____
_____
_____

Discussion of critical appraisal criteria

How to cite: Lockwood C, Munn Z, Porritt K. Qualitative research synthesis: methodological guidance for systematic reviewers utilizing meta-aggregation. Int J Evid Based Healthc. 2015;13(3):179–187.

1 **Congruity between the stated philosophical perspective and the research methodology**
Does the report clearly state the philosophical or theoretical premises on which the study is based? Does the report clearly state the methodological approach adopted on which the study is based? Is there congruence between the two? For example:

A report may state that the study adopted a critical perspective and participatory action research methodology was followed. Here there is congruence between a critical view (focusing on knowledge arising out of critique, action and reflection) and action research (an approach that focuses on firstly working with groups to reflect on issues or practices, then considering how they could be different; then acting to create a change; and finally identifying new knowledge arising out of the action taken). However, a report may state that the study adopted an interpretive perspective and used survey methodology. Here there is incongruence between an interpretive view (focusing on knowledge arising out of studying what phenomena mean to individuals or groups) and surveys (an approach that focuses on asking standard questions to a defined study population); a report may state that the study was qualitative or used qualitative methodology (such statements do not demonstrate rigour in design) or make no statement on philosophical orientation or methodology.

2 **Congruity between the research methodology and the research question or objectives**
Is the study methodology appropriate for addressing the research question? For example:

A report may state that the research question was to seek understandings of the meaning of pain in a group of people with rheumatoid arthritis and that a phenomenological approach was taken. Here, there is congruity between this question and the methodology. A report may state that the research question was to establish the effects of counselling on the severity of pain experience and that an ethnographic approach was pursued. A question that tries to establish cause and effect cannot be addressed by using an ethnographic approach (as ethnography sets out to develop understandings of cultural practices) and thus, this would be incongruent.

3 **Congruity between the research methodology and the methods used to collect data**
Are the data collection methods appropriate to the methodology? For example:

A report may state that the study pursued a phenomenological approach and data was collected through phenomenological interviews. There is congruence between the methodology and data collection; a report may state that the study pursued a phenomenological approach and data was collected through a postal questionnaire. There is incongruence between the methodology and data collection here as phenomenology seeks to elicit rich descriptions of the experience of a phenomena that cannot be achieved through seeking written responses to standardised questions.

4 **Congruity between the research methodology and the representation and analysis of data**
Are the data analysed and represented in ways that are congruent with the stated methodological position? For example:

A report may state that the study pursued a phenomenological approach to explore people's experience of grief by asking participants to describe their experiences of grief. If the text generated from asking these questions is searched to establish the meaning of grief to participants, and the meanings of all participants are included in the report findings, then this represents congruity; the same report may, however, focus only on those meanings that were common to all participants and discard single reported meanings. This would not be appropriate in phenomenological work.

5 There is congruence between the research methodology and the interpretation of results

Are the results interpreted in ways that are appropriate to the methodology? For example:

A report may state that the study pursued a phenomenological approach to explore people's experience of facial disfigurement and the results are used to inform practitioners about accommodating individual differences in care. There is congruence between the methodology and this approach to interpretation; a report may state that the study pursued a phenomenological approach to explore people's experience of facial disfigurement and the results are used to generate practice checklists for assessment. There is incongruence between the methodology and this approach to interpretation as phenomenology seeks to understand the meaning of a phenomenon for the study participants and cannot be interpreted to suggest that this can be generalised to total populations to a degree where standardised assessments will have relevance across a population.

6 Locating the researcher culturally or theoretically

Are the beliefs and values, and their potential influence on the study declared? For example:

The researcher plays a substantial role in the qualitative research process and it is important, in appraising evidence that is generated in this way, to know the researcher's cultural and theoretical orientation. A high-quality report will include a statement that clarifies this.

7 Influence of the researcher on the research, and vice-versa, is addressed

Is the potential for the researcher to influence the study and for the potential of the research process itself to influence the researcher and her/his interpretations acknowledged and addressed? For example:

Is the relationship between the researcher and the study participants addressed? Does the researcher critically examine her/his own role and potential influence during data collection? Is it reported how the researcher responded to events that arose during the study?

8 Representation of participants and their voices

Generally, reports should provide illustrations from the data to show the basis of their conclusions and to ensure that participants are represented in the report.

9 Ethical approval by an appropriate body

A statement on the ethical approval process followed should be in the report.

10 Relationship of conclusions to analysis, or interpretation of the data

This criterion concerns the relationship between the findings reported and the views or words of study participants. In appraising a paper, appraisers seek to satisfy themselves that the conclusions drawn by the research are based on the data collected; data being the text generated through observation, interviews or other processes.

Checklist for quasi-experimental studies (non-randomised experimental studies)

Introduction

JBI is an international research organisation based in the Faculty of Health and Medical Sciences at the University of Adelaide, South Australia. JBI develops and delivers unique evidence-based information, software, education and training designed to improve healthcare practice and health outcomes. With over 70 Collaborating Entities, servicing over 90 countries, JBI is a recognised global leader in evidence-based healthcare.

JBI systematic reviews

The core of evidence synthesis is the systematic review of literature of a particular intervention, condition or issue. The systematic review is essentially an analysis of the available literature (i.e., evidence) and a judgment of the effectiveness or otherwise of a practice, involving a series of complex steps. JBI takes a particular view on what counts as evidence and the methods utilised to synthesise those different types of evidence. In line with this broader view of evidence, JBI has developed theories, methodologies and rigorous processes for the critical appraisal and synthesis of these diverse forms of evidence in order to aid in clinical decision-making in healthcare. There now exists JBI guidance for conducting reviews of effectiveness research, qualitative research, prevalence/incidence, aetiology/risk, economic evaluations, text/opinion, diagnostic test accuracy, mixed-methods, umbrella reviews and scoping reviews. Further information regarding JBI systematic reviews can be found in the JBI Evidence Synthesis Manual.

JBI critical appraisal tools

All systematic reviews incorporate a process of critique or appraisal of the research evidence. The purpose of this appraisal is to assess the methodological quality of a study and to determine the extent to which a study has addressed the possibility of bias in its design, conduct and analysis. All papers selected for inclusion in the systematic review (i.e., those that meet the inclusion criteria described in the protocol) need to be subjected to rigorous appraisal by two critical appraisers. The results of this appraisal can then be used to inform synthesis and interpretation of the results of the study. JBI critical appraisal tools have been developed by the JBI and collaborators and approved by the JBI Scientific Committee following extensive peer review. Although designed for use in systematic reviews, JBI critical appraisal tools can also be used when creating Critically Appraised Topics (CAT), in journal clubs and as an educational tool.

JBI critical appraisal checklist for quasi-experimental studies

Reviewer _____ Date _____

Author _____ Year _____ Record Number _____

	Yes	No	Unclear	Not applicable
1 Is it clear in the study what is the 'cause' and what is the 'effect' (i.e. there is no confusion about which variable comes first)?	□	□	□	□
2 Were the participants included in any comparisons similar?	□	□	□	□
3 Were the participants included in any comparisons receiving similar treatment/care, other than the exposure or intervention of interest?	□	□	□	□
4 Was there a control group?	□	□	□	□
5 Were there multiple measurements of the outcome both pre and post the intervention/exposure?	□	□	□	□
6 Was follow up complete and if not, were differences between groups in terms of their follow up adequately described and analysed?	□	□	□	□
7 Were the outcomes of participants included in any comparisons measured in the same way?	□	□	□	□
8 Were outcomes measured in a reliable way?	□	□	□	□
9 Was appropriate statistical analysis used?	□	□	□	□

Overall appraisal: Include □   Exclude □   Seek further info □

Comments (Including reason for exclusion)

_____
_____
_____

Explanation for the critical appraisal tool for quasi-experimental studies

How to cite: Tufanaru C, Munn Z, Aromataris E, Campbell J, Hopp L. Chapter 3: Systematic reviews of effectiveness. In: Aromataris E, Munn Z (Editors). JBI Manual for Evidence Synthesis. JBI, 2020. Available from https://synthesismanual.jbi.global

Critical appraisal tool for Quasi-experimental studies (experimental studies without random allocation)

Answers: Yes, No, Unclear or Not/Applicable

1 Is it clear in the study what is the 'cause' and what is the 'effect' (i.e. there is no confusion about which variable comes first)?
Ambiguity with regards to the temporal relationship of variables constitutes a threat to the internal validity of a study exploring causal relationships. The 'cause' (the independent variable, that is, the treatment or intervention of interest) should occur in time before the explored 'effect' (the dependent variable, which is the effect or outcome of interest). Check if it is clear which variable is manipulated as a potential cause. Check if it is clear which variable is measured as the effect of the potential cause. Is it clear that the 'cause' was manipulated before the occurrence of the 'effect'?

2 **Were the participants included in any comparisons similar?**

The differences between participants included in compared groups constitute a threat to the internal validity of a study exploring causal relationships. If there are differences between participants included in compared groups there is a risk of selection bias. If there are differences between participants included in the compared groups maybe the 'effect' cannot be attributed to the potential 'cause', as maybe it is plausible that the 'effect' may be explained by the differences between participants, that is, by selection bias. Check the characteristics reported for participants. Are the participants from the compared groups similar with regards to the characteristics that may explain the effect even in the absence of the 'cause', for example, age, severity of the disease, stage of the disease, co-existing conditions and so on? [NOTE: In one single group pre-test/post-test studies where the patients are the same (the same one group) in any pre-post comparisons, the answer to this question should be 'yes'.]

3 **Were the participants included in any comparisons receiving similar treatment/care, other than the exposure or intervention of interest?**

In order to attribute the 'effect' to the 'cause' (the exposure or intervention of interest), assuming that there is no selection bias, there should be no other difference between the groups in terms of treatments or care received, other than the manipulated 'cause' (the intervention of interest). If there are other exposures or treatments occurring in the same time with the 'cause', other than the intervention of interest, then potentially the 'effect' cannot be attributed to the intervention of interest, as it is plausible that the 'effect' may be explained by other exposures or treatments, other than the intervention of interest, occurring in the same time with the intervention of interest. Check the reported exposures or interventions received by the compared groups. Are there other exposures or treatments occurring in the same time with the intervention of interest? Is it plausible that the 'effect' may be explained by other exposures or treatments occurring in the same time with the intervention of interest?

4 **Was there a control group?**

Control groups offer the conditions to explore what would have happened with groups exposed to other different treatments, other than to the potential 'cause' (the intervention of interest). The comparison of the treated group (the group exposed to the examined 'cause', that is, the group receiving the intervention of interest) with such other groups strengthens the examination of the causal plausibility. The validity of causal inferences is strengthened in studies with at least one independent control group compared to studies without an independent control group. Check if there are independent, separate groups, used as control groups in the study. [Note: The control group should be an independent, separate control group, not the pre-test group in a single group pre-test post-test design.]

5 **Were there multiple measurements of the outcome both pre and post the intervention/ exposure?**

In order to show that there is a change in the outcome (the 'effect') as a result of the intervention/treatment (the 'cause') it is necessary to compare the results of measurement before and after the intervention/treatment. If there is no measurement before the treatment and only measurement after the treatment is available it is not known if there is a change after the treatment compared to before the treatment. If multiple measurements are collected before the intervention/treatment is implemented then it is possible to explore the plausibility of alternative explanations other than the proposed 'cause' (the intervention of interest) for the observed 'effect', such as the naturally occurring changes in the absence of the 'cause', and changes of high (or low) scores towards less extreme values even in the absence of the 'cause' (sometimes called regression to the mean). If multiple measurements are collected after the intervention/treatment is implemented it is possible to explore the changes of the 'effect' in time in each group and to compare these changes across the groups. Check if measurements were collected before the intervention of interest was implemented. Were there multiple pre-test measurements? Check if measurements were collected after the intervention of interest was implemented. Were there multiple post-test measurements?

6   **Was follow up complete and if not, were differences between groups in terms of their follow up adequately described and analysed?**
If there are differences with regards to the loss to follow up between the compared groups these differences represent a threat to the internal validity of a study exploring causal effects as these differences may provide a plausible alternative explanation for the observed 'effect' even in the absence of the 'cause' (the treatment or exposure of interest). Check if there were differences with regards to the loss to follow up between the compared groups. If follow up was incomplete (i.e., there is incomplete information on all participants), examine the reported details about the strategies used in order to address incomplete follow up, such as descriptions of loss to follow up (absolute numbers; proportions; reasons for loss to follow up; patterns of loss to follow up) and impact analyses (the analyses of the impact of loss to follow up on results). Was there a description of the incomplete follow up (number of participants and the specific reasons for loss to follow up)? If there are differences between groups with regards to the loss to follow up, was there an analysis of patterns of loss to follow up? If there are differences between the groups with regards to the loss to follow up, was there an analysis of the impact of the loss to follow up on the results?

7   **Were the outcomes of participants included in any comparisons measured in the same way?**
If the outcome (the 'effect') is not measured in the same way in the compared groups there is a threat to the internal validity of a study exploring a causal relationship as the differences in outcome measurements may be confused with an effect of the treatment or intervention of interest (the 'cause'). Check if the outcomes were measured in the same way. Same instrument or scale used? Same measurement timing? Same measurement procedures and instructions?

8   **Were outcomes measured in a reliable way?**
Unreliability of outcome measurements is one threat that weakens the validity of inferences about the statistical relationship between the 'cause' and the 'effect' estimated in a study exploring causal effects. Unreliability of outcome measurements is one of different plausible explanations for errors of statistical inference with regards to the existence and the magnitude of the effect determined by the treatment ('cause'). Check the details about the reliability of measurement such as the number of raters, training of raters, the intra-rater reliability, and the inter-raters reliability within the study (not to external sources). This question is about the reliability of the measurement performed in the study, it is not about the validity of the measurement instruments/scales used in the study. *[Note: Two other important threats that weaken the validity of inferences about the statistical relationship between the* 'cause' *and the* 'effect' *are low statistical power and the violation of the assumptions of statistical tests. These other threats are not explored within Question 8, these are explored within Question 9.]*

9 **Was appropriate statistical analysis used?**

Inappropriate statistical analysis may cause errors of statistical inference with regards to the existence and the magnitude of the effect determined by the treatment ('cause'). Low statistical power and the violation of the assumptions of statistical tests are two important threats that weakens the validity of inferences about the statistical relationship between the 'cause' and the 'effect'. Check the following aspects: if the assumptions of statistical tests were respected; if appropriate statistical power analysis was performed; if appropriate effect sizes were used; if appropriate statistical procedures or methods were used given the number and type of dependent and independent variables, the number of study groups, the nature of the relationship between the groups (independent or dependent groups), and the objectives of statistical analysis (association between variables; prediction; survival analysis, etc.).

JBI critical appraisal tool for assessment of risk of bias for randomised controlled trials

Introduction

JBI is a global organisation promoting and supporting evidence-based decisions that improve health and health service delivery.

JBI offers a unique range of solutions to access, appraise and apply the best available evidence.

JBI's approach to evidence-based healthcare is unique. JBI considers evidence-based healthcare as decision making that considers the feasibility, appropriateness, meaningfulness and effectiveness (FAME) of healthcare practice.

JBI systematic reviews

The core of evidence synthesis is the systematic review of literature of a particular intervention, condition or issue. The systematic review is essentially an analysis of the available evidence and a judgment of the effectiveness or otherwise of a practice, involving a series of complex steps. JBI take a particular view on what counts as evidence and the methods utilised to synthesise those different types of evidence. In line with this broader view of evidence, JBI has developed theories, methodologies and rigorous processes for the critical appraisal and synthesis of these diverse forms of evidence in order to aid in clinical decision-making in health care. Guidance now exists for conducting reviews of effectiveness research, qualitative research, prevalence/incidence, aetiology/risk, economic evaluations, text/opinion, diagnostic test accuracy, mixed-methods, umbrella reviews and scoping reviews. Further information regarding JBI systematic reviews can be found in the JBI Manual for Evidence Synthesis.

JBI critical appraisal tools

All systematic reviews incorporate a process of critique or appraisal of the research evidence. The purpose of this appraisal for quantitative evidence is to determine the extent to which a study has addressed the possibility of bias in its design, conduct and analysis. All papers selected for inclusion in the systematic review (i.e., those that meet the inclusion criteria described in the protocol) need to be subjected to rigorous appraisal by two critical appraisers. The results of this appraisal can then be used to inform synthesis and interpretation of the results of the study. Although designed for use in systematic reviews, JBI critical appraisal tools can also be used when creating Critically Appraised Topics (CATs), in journal clubs and as an educational tool.

How were these tools developed?

JBI critical appraisal tools have been developed by JBI and collaborators. The particular iteration of this tool was developed by the JBI Effectiveness Methods Group following oversight by the JBI Scientific Committee.

Like the previous versions of these tools, this version presents signalling questions to prompt reviewers to identify whether certain safeguards of bias have been met, in the primary literature under review. However, unlike previous iterations of this tool, this version has separated questions into whether they provide an answer relating to internal, external or statistical conclusion validity. For questions related to internal validity, these have been further separated to identify what domain of bias they are referring. Finally, this tool has also been structured to facilitate judgments related to bias at different levels (e.g. bias at the outcome level or bias at the result level) where appropriate.

These tools have been approved following extensive peer review by the JBI Scientific Committee.

How to cite: Barker TH, Stone JC, Sears K, Klugar M, Tufanaru C, Leonardi-Bee J, Aromataris E, Munn Z. The revised JBI critical appraisal tool for the assessment of risk of bias for randomized controlled trials. JBI Evidence Synthesis. 2023;21(3):494–506

Question guidance

How to use the JBI Tools for the assessment of risk of bias

Each question presented in a JBI tool for the assessment of risk of bias for quantitative study designs answers a question related to certain *categories of validity* and *domains of bias*. The concept of validity is often used when referring to the soundness or rigour in which a study was conducted, and whether the results of the study are likely to be true and generalisable. At JBI we have broken this down to include three separate categories that constitute *validity*; these include internal validity, external validity, statistical conclusion validity. In addition, we have also included comprehensiveness of reporting.

Questions categorised as "Internal Validity" are then further organised to specific domains of bias in which they relate. The domains of bias that are used as an indicator of internal validity include bias related to selection and allocation, bias related to administration of the intervention/exposure, bias related to assessment, detection and measurement of the outcome, bias related to participant retention, bias related to temporal precedence, bias related to classification of the exposure, bias related to confounding factors and bias related to selective reporting and/or publication bias.

For more information, please see Barker et al. 2022

Table 1 The JBI critical appraisal tool for RCTs

Assessor: Date of Appraisal: Record Number:

Study Author: Study Title: Study Year:

	Choice – Comments/Justification	Yes	No	Unclear	N/A
Internal Validity					
Bias related to selection and allocation					
1 Was true randomisation used for assignment of participants to treatment groups?		☐	☐	☐	☐
2 Was allocation to treatment groups concealed?		☐	☐	☐	☐
3 Were treatment groups similar at the baseline?		☐	☐	☐	☐
Bias related to administration of intervention/exposure					
4 Were participants blind to treatment assignment?		☐	☐	☐	☐
5 Were those delivering the treatment blind to treatment assignment?		☐	☐	☐	☐
6 Were treatment groups treated identically other than the intervention of interest?		☐	☐	☐	☐
Bias related to assessment, detection and measurement of the outcome					
7 Were outcome assessors blind to treatment assignment?		**Yes**	**No**	**Unclear**	**N/A**
Outcome 1		☐	☐	☐	☐
Outcome 2		☐	☐	☐	☐
Outcome 3		☐	☐	☐	☐
Outcome 4		☐	☐	☐	☐
Outcome 5		☐	☐	☐	☐
Outcome 6		☐	☐	☐	☐
Outcome 7		☐	☐	☐	☐
8 Were outcomes measured in the same way for treatment groups?		**Yes**	**No**	**Unclear**	**N/A**
Outcome 1		☐	☐	☐	☐
Outcome 2		☐	☐	☐	☐
Outcome 3		☐	☐	☐	☐
Outcome 4		☐	☐	☐	☐
Outcome 5		☐	☐	☐	☐
Outcome 6		☐	☐	☐	☐
Outcome 7		☐	☐	☐	☐

	Yes	No	Unclear	N/A
9 Were outcomes measured in a reliable way				
Outcome 1	☐	☐	☐	☐
Outcome 2	☐	☐	☐	☐
Outcome 3	☐	☐	☐	☐
Outcome 4	☐	☐	☐	☐
Outcome 5	☐	☐	☐	☐
Outcome 6	☐	☐	☐	☐
Outcome 7	☐	☐	☐	☐

Bias related to participant retention

10 Was follow up complete and if not, were differences between groups in terms of their follow up adequately described and analysed?

	Yes	No	Unclear	N/A
Outcome 1	☐	☐	☐	☐
Result 1	☐			
Result 2	☐			
Result 3	☐			
Outcome 2	☐	☐	☐	☐
Result 1	☐			
Result 2	☐			
Result 3	☐			
Outcome 3	☐	☐	☐	☐
Result 1	☐			
Result 2	☐			
Result 3	☐			
Outcome 4	☐	☐	☐	☐
Result 1	☐			
Result 2	☐			
Result 3	☐			
Outcome 5	☐	☐	☐	☐
Result 1	☐			
Result 2	☐			
Result 3	☐			

(Continued)

Table 1 (Continued)

Assessor: Date of Appraisal: Record Number:

Study Author: Study Title: Study Year:

	Yes	No	Unclear	N/A
Outcome 6	☐	☐	☐	☐
Result 1	☐	☐	☐	☐
Result 2	☐	☐	☐	☐
Result 3	☐	☐	☐	☐
Outcome 7	☐	☐	☐	☐
Result 1	☐	☐	☐	☐
Result 2	☐	☐	☐	☐
Result 3				
Statistical Conclusion Validity				
11 Were participants analysed in the groups to which they were randomised?				
Outcome 1	☐	☐	☐	☐
Result 1	☐	☐	☐	☐
Result 2	☐	☐	☐	☐
Result 3	☐	☐	☐	☐
Outcome 2	☐	☐	☐	☐
Result 1	☐	☐	☐	☐
Result 2	☐	☐	☐	☐
Result 3	☐	☐	☐	☐
Outcome 3	☐	☐	☐	☐
Result 1	☐	☐	☐	☐
Result 2	☐	☐	☐	☐
Result 3	☐	☐	☐	☐
Outcome 4	☐	☐	☐	☐
Result 1	☐	☐	☐	☐
Result 2	☐	☐	☐	☐
Result 3	☐	☐	☐	☐

	Yes	No	Unclear	N/A
Outcome 5				
Result 1	☐	☐	☐	☐
Result 2	☐	☐	☐	☐
Result 3	☐	☐	☐	☐
Outcome 6				
Result 1	☐	☐	☐	☐
Result 2	☐	☐	☐	☐
Result 3	☐	☐	☐	☐
Outcome 7				
Result 1	☐	☐	☐	☐
Result 2	☐	☐	☐	☐
Result 3	☐	☐	☐	☐

12 Was appropriate statistical analysis used?

	Yes	No	Unclear	N/A
Outcome 1				
Result 1	☐	☐	☐	☐
Result 2	☐	☐	☐	☐
Result 3	☐	☐	☐	☐
Outcome 2				
Result 1	☐	☐	☐	☐
Result 2	☐	☐	☐	☐
Result 3	☐	☐	☐	☐
Outcome 3				
Result 1	☐	☐	☐	☐
Result 2	☐	☐	☐	☐
Result 3	☐	☐	☐	☐
Outcome 4				
Result 1	☐	☐	☐	☐
Result 2	☐	☐	☐	☐
Result 3	☐	☐	☐	☐
Outcome 5				
Result 1	☐	☐	☐	☐
Result 2	☐	☐	☐	☐
Result 3	☐	☐	☐	☐

(Continued)

Table 1 (Continued)

Assessor: Date of Appraisal: Record Number:

Study Author: Study Title: Study Year:

	Yes	No	Unclear	N/A
Outcome 6	☐	☐	☐	☐
Result 1	☐	☐	☐	☐
Result 2	☐	☐	☐	☐
Result 3	☐	☐	☐	☐
Outcome 7	☐	☐	☐	☐
Result 1	☐	☐	☐	☐
Result 2	☐	☐	☐	☐
Result 3	☐	☐	☐	☐
13 Was the trial design appropriate and any deviations from the standard RCT design (individual randomisation, parallel groups) accounted for in the conduct and analysis of the trial?	☐	☐	☐	☐

Overall appraisal: **Include:** ☐   **Exclude:** ☐   **Seek Further Info:** ☐

Comments:

Question 1: Was true randomisation used for assignment of participants to treatment groups?

Category: Internal validity
Domain: Bias related to selection and allocation
Appraisal: Study level

If participants are not allocated to treatment and control groups by random assignment there is a risk that this assignment to groups can be influenced by the known characteristics of the participants themselves. These known characteristics of the participants may distort the comparability of the groups (i.e. does the intervention group contain more people over the age of 65 as compared to the control?). A true random assignment of participants to the groups means that a procedure is used that allocates the participants to groups purely based on chance, not influenced by any known characteristics of the participants. Reviewers should check the details about the randomisation procedure used for allocation of the participants to study groups. Was a true chance (random) procedure used? For example, was a list of random numbers used? Was a computer-generated list of random numbers used? Was a statistician, external to the research team consulted for the randomisation sequence generation? Additionally, reviewers should check that the authors are not stating they have used random approaches when they have instead used systematic approaches (such as allocating by days of the week).

Question 2: Was allocation to groups concealed?

Category: Internal validity
Domain: Bias related to selection and allocation
Appraisal: Study level

If those allocating participants to the compared groups are aware of which group is next in the allocation process (i.e., the treatment or control group), there is a risk that they may deliberately and purposefully intervene in the allocation of patients. This may result in the preferential allocation of patients to the treatment group or to the control group. This may directly distort the results of the study, as participants no longer have an equal and random chance to belong to each group compared. Concealment of allocation refers to procedures that prevent those allocating patients from knowing before allocation which treatment or control is next in the allocation process. Reviewers should check the details about the procedure used for allocation concealment. Was an appropriate allocation concealment procedure used? For example, was central randomisation used? Were sequentially numbered, opaque and sealed envelopes used? Were coded drug packs used?

Question 3: Were treatment groups similar at the baseline?

Category: Internal validity
Domain: Bias related to selection and allocation
Appraisal: Study level

As with Question 1, any differences between the known characteristics of participants included in compared groups constitutes a threat to internal validity. If differences in these characteristics do exist, then there is potential that the 'effect' cannot be attributed to the potential 'cause' (the examined intervention or treatment). This is because the 'effect' may be explained by the differences between participant characteristics and not due to the intervention/treatment of interest. Reviewers should check the characteristics reported for participants. Are the participants from the compared groups similar with regards to the characteristics that may explain the effect even in the absence of the 'cause', for example, age, severity of the disease, stage of the disease, co-existing conditions and so on? Reviewers should check the proportions of participants with specific relevant characteristics in the compared groups. [Note: **Do NOT** only consider the P-value for the statistical testing of the differences between groups with regards to the baseline characteristics.]

Question 4: Were participants blind to treatment assignment?

Category: Internal validity
Domain: Bias related to administration of intervention/exposure
Appraisal: Study level

Participants that are aware of their allocation to either the treatment or the control may behave, respond, or react differently to their assigned treatment (or control) than compared to participants that remain unaware of their allocation. Blinding of participants is a technique used to minimise this risk. Blinding refers to procedures that prevent participants from knowing which group they are allocated. If blinding has been followed, participants are not aware if they are in the group receiving the treatment of interest or if they are in any other group receiving the control interventions. Reviewers should check the details reported in the article about the blinding of participants with regards to treatment assignment. Was an appropriate blinding procedure used? For example, were identical capsules or syringes used? Were identical devices used? Be aware of different terms used, blinding is sometimes also called masking.

Question 5: Were those delivering the treatment blind to treatment assignment?

Category: Internal validity
Domain: Bias related to administration of intervention/exposure
Appraisal: Study level

Like Question 4, those delivering the treatment that are aware of participant allocation to either treatment or control, may treat participants differently than compared to those that remain unaware of participant allocation. There is the risk that any potential change in behaviour may influence the implementation of the compared treatments and the results of the study may be distorted. Blinding of those delivering treatment is used to minimise this risk. When this level of blinding has been achieved, those delivering the treatment are not aware if they are treating the group receiving the treatment of interest or if they are treating any other group receiving the control interventions. Reviewers should check the details reported in the article about the blinding of those delivering treatment with regards to treatment assignment. Is there any information in the article about those delivering the treatment? Were those delivering the treatment unaware of the assignments of participants to the compared groups?

Question 6: Were treatment groups treated identically other than the intervention of interest?

Category: Internal validity
Domain: Bias related to administration of intervention/exposure
Appraisal: Study level

To attribute the 'effect' to the 'cause', (assuming no bias related to selection and allocation) there should be no other difference between the groups in terms of treatment or care received, other than the treatment or intervention controlled by the researchers. If there are other exposures or treatments occurring at the same time with the 'cause' (the treatment or intervention of interest), then the 'effect' can potentially not be attributed to the examined 'cause' (the investigated treatment). This is because it is plausible that the 'effect' may be explained by these other exposures or treatments that occurred at the same time with the 'cause'. Reviewers should check the reported exposures or interventions received by the compared groups. Are there other exposures or treatments occurring at the same time with the 'cause'? Is it plausible that the 'effect' may be explained by other exposures or treatments occurring at the same time with the 'cause'? Is it clear that there is no other difference between the groups in terms of treatment or care received, other than the treatment or intervention of interest?

Question 7: Were outcome assessors blind to treatment assignment?

Category: Internal validity
Domain: Bias related to assessment, detection and measurement of the outcome
Appraisal: Outcome level

Like Questions 4 and 5, those assessing the outcomes that are aware of participant allocation to either treatment or control, may treat participants differently than compared to those that remain unaware of participant allocation. Therefore, there is a risk that the measurement of the outcomes between groups may be distorted, and the results of the study may themselves be distorted. Blinding of outcomes assessors is used in order to minimise this risk. Reviewers should check the details reported in the article about the blinding of outcomes assessors with regards to treatment assignment. Is there any information in the article about outcomes assessors? Were those assessing the treatment's effects on outcomes unaware of the assignments of participants to the compared groups?

Question 8: Were outcomes measured in the same way for treatment groups?

Category: Internal validity
Domain: Bias related to assessment, detection and measurement of the outcome
Appraisal: Outcome level

If the outcome is not measured in the same way in the compared groups, there is a threat to the internal validity of a study. Any differences in outcome measurements may be due to the method of measurement employed between the two groups, and not due to the intervention/treatment of interest. Reviewers should check if the outcomes were measured in the same way. Same instrument or scale used? Same measurement timing? Same measurement procedures and instructions?

Question 9: Were outcomes measured in a reliable way?

Category: Internal validity
Domain: Bias related to assessment, detection and measurement of the outcome
Appraisal: Outcome level

Unreliability of outcome measurements is one threat that weakens the validity of inferences about the statistical relationship between the 'cause' and the 'effect' estimated in a study exploring causal effects. Unreliability of outcome measurements is one of the different plausible explanations for errors of statistical inference with regards to the existence and the magnitude of the effect determined by the treatment ('cause'). Reviewers should check the details about the reliability of the measurement used, such as the number of raters, training of raters, the intra-rater and the inter-raters reliability within the study (not as reported in external sources). This question is about the reliability of the measurement performed in the study, it is not about the validity of the measurement instruments/scales used in the study. Finally, some outcomes may not rely on instruments or scales (e.g. death) and reliability of the measurements may need to be assessed in the context of the study being reviewed. [Note: Two other important threats that weaken the validity of inferences about the statistical relationship between the 'cause' and the 'effect' are low statistical power and the violation of the assumptions of statistical tests. These other two threats are explored within Question 12).]

Question 10: Was follow up complete and if not, were differences between groups in terms of their follow up adequately described and analysed?

Category: Internal validity
Domain: Bias related to participant retention
Appraisal: Result level

For this question, follow up refers to the period from the moment of randomisation to any point in which the groups are compared during the trial. This question asks if there is complete knowledge (measurements, observations, etc.) for the entire duration of the trial for all randomly allocated participants. If there is incomplete follow up from all randomly allocated participants, this is known as post-assignment attrition. As RCTs are not perfect, there is almost always post-assignment attrition, and the focus of this question is on the appropriate exploration of post-assignment attrition. If differences do exist with regards to the post-assignment attrition between the compared groups of an RCT, then there is a threat to the internal validity of that study. This is because these differences may provide a plausible alternative explanation for the observed 'effect' even in the absence of the 'cause' (the treatment or intervention of interest). It is important to note that with regards post-assignment attrition, it is not enough to know the number of participants and the proportions of participants with incomplete data; the reasons for loss to follow up are essential in the analysis of risk of bias.

Reviewers should check if there were differences with regards to the loss to follow up between the compared groups. If follow up was incomplete (incomplete information on all participants), examine the reported details about the strategies used to address incomplete follow up. This can include descriptions of loss to follow up (absolute numbers; proportions; reasons for loss to follow up) and impact analyses (the analyses of the impact of loss to follow up on results). Was there a description of the incomplete follow up including the number of participants and the specific reasons for loss to follow up? Even if follow up was incomplete, but balanced between groups, if the reasons for loss to follow up are different (e.g., side effects caused by the intervention of interest), these may impose a risk of bias if not appropriately explored in the analysis. If there are differences between groups with regards to the loss to follow up (numbers/proportions and reasons), was there an analysis of patterns of loss to follow up? If there are differences between the groups with regards to the loss to follow up, was there an analysis of the impact of the loss to follow up on the results? [Note: Question 10 is NOT about intention-to-treat (ITT) analysis; Question 11 is about ITT analysis.]

Question 11: Were participants analysed in the groups to which they were randomised?

Category: Statistical conclusion validity
Appraisal: Result level

This question is about the intention-to-treat (ITT) analysis. There are different statistical analysis strategies available for the analysis of data from RCTs, such as intention-to-treat analysis (known also as intent to treat; abbreviated, ITT), per-protocol analysis, and as-treated analysis. In the ITT analysis the participants are analysed in the groups to which they were randomised. This means that regardless of whether participants received the intervention or control as assigned, were complaint with their planned assignment or participated for the entire study duration, they are still included in the analysis. The ITT analysis compares the outcomes for participants from the initial groups created by the initial random allocation of participants to those groups. Reviewers should check if an ITT analysis was reported; check the details of the ITT. Were participants analysed in the groups to which they were initially randomised, regardless of whether they participated in those groups, and regardless of whether they received the planned interventions?

[Note: The ITT analysis is a type of statistical analysis recommended in the Consolidated Standards of Reporting Trials (CONSORT) statement on best practices in trials reporting, and it is considered a marker of good methodological quality of the analysis of results of a randomised trial. The ITT is estimating the effect of offering the intervention, that is, the effect of instructing the participants to use or take the intervention; the ITT it is not estimating the effect of receiving the intervention of interest.]

Question 12: Was appropriate statistical analysis used?

Category: Statistical conclusion validity
Appraisal: Result level

Inappropriate statistical analysis may cause errors of statistical inference with regards to the existence and the magnitude of the effect determined by the treatment ('cause'). Low statistical power and the violation of the assumptions of statistical tests are two important threats that weaken the validity of inferences about the statistical relationship between the 'cause' and the 'effect'. Reviewers should check the following aspects: were the assumptions of the statistical tests were respected; if appropriate statistical power analysis was performed; if appropriate effect sizes were used; if appropriate statistical methods were used given the nature of the data and the objectives of statistical analysis (association between variables; prediction; survival analysis, etc.).

Question 13: Was the trial design appropriate and any deviations from the standard RCT design (individual randomisation, parallel groups) accounted for in the conduct and analysis of the trial?

Category: Statistical conclusion validity
Appraisal: Study level

The typical, parallel group RCT may not always be appropriate depending on the nature of the question being asked. Therefore, some additional RCT designs may have been employed that each come with their own additional considerations.

Crossover trials should only be conducted in people with a chronic, stable condition, where the intervention produces a short-term effect (i.e. relief in symptoms). Crossover trials should ensure there is an appropriate period of washout between treatments. This may also be considered under Question 6.

Cluster RCTs randomise groups individuals or groups (e.g., communities and wards), forming 'clusters'. When we are assessing outcomes on an individual level in cluster trials, there are unit-of-analysis issues, as individuals within a cluster are correlated. This should be considered by the study authors when conducting analysis, and ideally authors will report the intra-cluster correlation coefficient. This may also be considered under Question 12.

Stepped wedge RCTs may be appropriate to establish when and how a beneficial intervention may be best implemented within a defined setting, or due to logistical, practical, or financial considerations in the roll out of a new treatment/intervention. Data analysis in these trials should be conducted appropriately, considering the effects of time. This may also be considered under Question 12.

Checklist for systematic reviews and research synthesis

Introduction

JBI is an international research organisation based in the Faculty of Health and Medical Sciences at the University of Adelaide, South Australia. JBI develops and delivers unique evidence-based information, software, education and training designed to improve healthcare practice and health outcomes. With over 70 Collaborating Entities, servicing over 90 countries, JBI is a recognised global leader in evidence-based healthcare.

JBI systematic reviews

The core of evidence synthesis is the systematic review of literature of a particular intervention, condition or issue. The systematic review is essentially an analysis of the available literature (i.e., evidence) and a judgment of the effectiveness or otherwise of a practice, involving a series of complex steps. JBI takes a particular view on what counts as evidence and the methods utilised to synthesise those different types of evidence. In line with this broader view of evidence, JBI has developed theories, methodologies and rigorous processes for the critical appraisal and synthesis of these diverse forms of evidence in order to aid in clinical decision-making in healthcare. There now exists JBI guidance for conducting reviews of effectiveness research, qualitative research, prevalence/incidence, aetiology/risk, economic evaluations, text/opinion, diagnostic test accuracy, mixed-methods, umbrella reviews and scoping reviews. Further information regarding JBI systematic reviews can be found in the JBI Evidence Synthesis Manual.

JBI critical appraisal tools

All systematic reviews incorporate a process of critique or appraisal of the research evidence. The purpose of this appraisal is to assess the methodological quality of a study and to determine the extent to which a study has addressed the possibility of bias in its design, conduct and analysis. All papers selected for inclusion in the systematic review (i.e., those that meet the inclusion criteria described in the protocol) need to be subjected to rigorous appraisal by two critical appraisers. The results of this appraisal can then be used to inform synthesis and interpretation of the results of the study. JBI critical appraisal tools have been developed by the JBI and collaborators and approved by the JBI Scientific Committee following extensive peer review. Although designed for use in systematic reviews, JBI critical appraisal tools can also be used when creating Critically Appraised Topics (CAT), in journal clubs and as an educational tool.

JBI critical appraisal checklist for systematic reviews and research syntheses

Reviewer _____ Date _____

Author _____ Year _____ Record Number _____

	Yes	No	Unclear	Not applicable
1 Is the review question clearly and explicitly stated?	□	□	□	□
2 Were the inclusion criteria appropriate for the review question?	□	□	□	□
3 Was the search strategy appropriate?	□	□	□	□
4 Were the sources and resources used to search for studies adequate?	□	□	□	□
5 Were the criteria for appraising studies appropriate?	□	□	□	□
6 Was critical appraisal conducted by two or more reviewers independently?	□	□	□	□
7 Were there methods to minimise errors in data extraction?	□	□	□	□
8 Were the methods used to combine studies appropriate?	□	□	□	□
9 Was the likelihood of publication bias assessed?	□	□	□	□
10 Were recommendations for policy and/or practice supported by the reported data?	□	□	□	□
11 Were the specific directives for new research appropriate?	□	□	□	□

Overall appraisal: Include □   Exclude □   Seek further info □
Comments (Including reason for exclusion)

_____

_____

_____

JBI critical appraisal checklist for systematic reviews and research synthesis

How to cite: *Aromataris E, Fernandez R, Godfrey C, Holly C, Kahlil H, Tungpunkom P. Summarizing systematic reviews: methodological development, conduct and reporting of an Umbrella review approach. Int J Evid Based Healthc. 2015;13(3):132–40.*

When conducting an umbrella review using the JBI method, the critical appraisal instrument for Systematic Reviews should be used.

The primary and secondary reviewer should discuss each item in the appraisal instrument for each study included in their review. In particular, discussions should focus on what is considered acceptable to the aims of the review in terms of the specific study characteristics. When appraising systematic reviews this discussion may include issues such as what represents an adequate search strategy or appropriate methods of synthesis. The reviewers should be clear on what constitutes acceptable levels of information to allocate a positive appraisal compared with a negative, or response of "unclear". This discussion should ideally take place before the reviewers independently conduct the appraisal.

Within umbrella reviews, quantitative or qualitative systematic reviews may be incorporated, as well as meta-analyses of existing research. There are 11 questions to guide the appraisal of systematic reviews or meta-analyses. Each question should be answered as "yes", "no", or "unclear". Not applicable "NA" is also provided as an option and may be appropriate in rare instances.

1 **Is the review question clearly and explicitly stated?**

The review question is an essential step in the systematic review process. A well-articulated question defines the scope of the review and aids in the development of the search strategy to locate the relevant evidence. An explicitly stated question, formulated around its PICO (Population, Intervention, Comparator, Outcome) elements aids both the review team in the conduct of the review and the reader in determining if the review has achieved its objectives. Ideally the review question should be articulated in a published protocol; however, this will not always be the case with many reviews that are located.

2 **Were the inclusion criteria appropriate for the review question?**

The inclusion criteria should be identifiable from, and match the review question. The necessary elements of the PICO should be explicit and clearly defined. The inclusion criteria should be detailed and the included reviews should clearly be eligible when matched against the stated inclusion criteria. Appraisers of meta-analyses will find that inclusion criteria may include criteria around the ability to conduct statistical analyses which would not be the norm for a systematic review. The types of included studies should be relevant to the review question, for example, an umbrella review aiming to summarise a range of effective non-pharmacological interventions for aggressive behaviours amongst elderly patients with dementia will limit itself to including systematic reviews and meta-analyses that synthesise quantitative studies assessing the various interventions; qualitative or economic reviews would not be included.

3 **Was the search strategy appropriate?**

A systematic review should provide evidence of the search strategy that has been used to locate the evidence. This may be found in the methods section of the review report in some cases, or as an appendix that may be provided as supplementary information to the review publication. A systematic review should present a clear search strategy that addresses each of the identifiable PICO components of the review question. Some reviews may also provide a description of the approach to searching and how the terms that were ultimately used were derived, though due to limits on word counts in journals this may be more the norm in online only publications. There should be evidence of logical and relevant keywords and terms and also evidence that Subject Headings and Indexing terms have been used in the conduct of the search. Limits on the search should also be considered and their potential impact; for example, if a date limit was used, was this appropriate and/or justified? If only English language studies were included, will such a language bias have an impact on the review? The response to these considerations will depend, in part, on the review question.

4 **Were the sources and resources used to search for studies adequate?**

A systematic review should attempt to identify "all" the available evidence and as such there should be evidence of a comprehensive search strategy. Multiple electronic databases should be searched including major bibliographic citation databases such as MEDLINE and CINAHL. Ideally, other databases that are relevant to the review question should also be searched, for example, a systematic review with a question about a physical therapy intervention should also look to search the PEDro database, whilst a review focusing on an educational intervention should also search the ERIC. Reviews of effectiveness should aim to search trial registries. A comprehensive search is the ideal way to minimise publication bias, as a result, a well conducted systematic review should also attempt to search for grey literature, or "unpublished" studies; this may involve searching websites relevant to the review question, or thesis repositories.

5   Were the criteria for appraising studies appropriate?

The systematic review should present a clear statement that critical appraisal was conducted and provide the details of the items that were used to assess the included studies. This may be presented in the methods of the review, as an appendix of supplementary information, or as a reference to a source that can be located. The tools or instruments used should be appropriate for the review question asked and the type of research conducted. For example, a systematic review of effectiveness should present a tool or instrument that addresses aspects of validity for experimental studies and randomised controlled trials such as randomisation and blinding – if the review includes observational research to answer the same question a different tool would be more appropriate. Similarly, a review assessing diagnostic test accuracy may refer to the recognised QUADAS1 tool.

6   Was critical appraisal conducted by two or more reviewers independently?

Critical appraisal or some similar assessment of the quality of the literature included in a systematic review is essential. A key characteristic to minimise bias or systematic error in the conduct of a systematic review is to have the critical appraisal of the included studies completed independently and in duplicate by members of the review team. The systematic review should present a clear statement that critical appraisal was conducted by at least two reviewers working independently from each other and conferring where necessary to reach decision regarding study quality and eligibility on the basis of quality.

7   Were there methods to minimise errors in data extraction?

Efforts made by review authors during data extraction can also minimise bias or systematic errors in the conduct of a systematic review. Strategies to minimise bias may include conducting all data extraction in duplicate and independently, using specific tools or instruments to guide data extraction and some evidence of piloting or training around their use.

8   Were the methods used to combine studies appropriate?

A synthesis of the evidence is a key feature of a systematic review. The synthesis that is presented should be appropriate for the review question and the stated type of systematic review and evidence it refers to. If a meta-analysis has been conducted this needs to be reviewed carefully.

Was it appropriate to combine the studies? Have the reviewers assessed heterogeneity statistically and provided some explanation for heterogeneity that may be present? Often, where heterogeneous studies are included in the systematic review, narrative synthesis will be an appropriate method for presenting the results of multiple studies. If a qualitative review, are the methods that have been used to synthesise findings congruent with the stated methodology of the review? Is there adequate descriptive and explanatory information to support the final synthesised findings that have been constructed from the findings sourced from the original research?

9   Was the likelihood of publication bias assessed?

As mentioned, a comprehensive search strategy is the best means by which a review author may alleviate the impact of publication bias on the results of the review. Reviews may also present statistical tests such as Egger's test or funnel plots to also assess the potential presence of publication bias and its potential impact on the results of the review. This question will not be applicable to systematic reviews of qualitative evidence.

10 Were recommendations for policy and/or practice supported by the reported data?
Whilst the first nine (9) questions specifically look to identify potential bias in the conduct of a systematic review, the final questions are more indictors of review quality rather than validity. Ideally a review should present recommendations for policy and practice. Where these recommendations are made there should be a clear link to the results of the review. Is there evidence that the strength of the findings and the quality of the research been considered in the formulation of review recommendations?

11 Were the specific directives for new research appropriate?
The systematic review process is recognised for its ability to identify where gaps in the research, or knowledge base, around a particular topic exist. Most systematic review authors will provide some indication, often in the discussion section of the report, of where future research direction should lie. Where evidence is scarce or sample sizes that support overall estimates of effect are small and effect estimates are imprecise, repeating similar research to those identified by the review may be necessary and appropriate. In other instances, the case for new research questions to investigate the topic may be warranted.

Reference

1. Whiting, P., Rutjes, A. W. S., Reitsma, J. B., Bossuyt, P. M. M., & Kleijnen, J. (2003). The development of QUADAS: A tool for the quality assessment of studies of diagnostic accuracy included in systematic reviews. *BMC Medical Research Methodology, 3*, 25. https://doi.org/10.1186/1471-2288-3-25.

Checklist for textual evidence: expert opinion

Introduction

JBI is an international research organisation based in the Faculty of Health and Medical Sciences at the University of Adelaide, South Australia. JBI develops and delivers unique evidence-based information, software, education and training designed to improve healthcare practice and health outcomes. With over 70 Collaborating Entities, servicing over 90 countries, JBI is a recognised global leader in evidence-based healthcare.

JBI systematic reviews

The core of evidence synthesis is the systematic review of literature of a particular intervention, condition or issue. The systematic review is essentially an analysis of the available literature (i.e., evidence) and a judgment of the effectiveness or otherwise of a practice, involving a series of complex steps. JBI takes a particular view on what counts as evidence and the methods utilised to synthesise those different types of evidence. In line with this broader view of evidence, JBI has developed theories, methodologies and rigorous processes for the critical appraisal and synthesis of these diverse forms of evidence in order to aid in clinical decision-making in healthcare. There now exists JBI guidance for conducting reviews of effectiveness research, qualitative research, prevalence/incidence, aetiology/risk, economic evaluations, textual evidence, diagnostic test accuracy, mixed-methods, umbrella reviews and scoping reviews. Further information regarding JBI systematic reviews can be found in the JBI Evidence Synthesis Manual.

JBI critical appraisal tools

All systematic reviews incorporate a process of critique or appraisal of the research evidence. The purpose of this appraisal is to assess the methodological quality of a study and to determine the extent to which a study has addressed the possibility of bias in its design, conduct and analysis. All papers selected for inclusion in the systematic review (i.e., those that meet the inclusion criteria described in the protocol) need to be subjected to rigorous appraisal by two critical appraisers. The results of this appraisal can then be used to inform synthesis and interpretation of the results of the study. JBI critical appraisal tools have been developed by the JBI and collaborators and approved by the JBI Scientific Committee following extensive peer review. Although designed for use in systematic reviews, JBI critical appraisal tools can also be used when creating Critically Appraised Topics (CAT), in journal clubs and as an educational tool.

JBI critical appraisal checklist for textual evidence: expert opinion

Reviewer _____ Date _____

Author _____ Year _____ Record Number _____

	Yes	No	Unclear	Not applicable
1 Is the source of the opinion clearly identified?	☐	☐	☐	☐
2 Does the source of opinion have standing in the field of expertise?	☐	☐	☐	☐
3 Are the interests of the relevant population the central focus of the opinion?	☐	☐	☐	☐
4 Does the opinion demonstrate a logically defended argument to support the conclusions drawn?	☐	☐	☐	☐
5 Is there reference to the extant literature?	☐	☐	☐	☐
6 Is any incongruence with the literature/sources logically defended?	☐	☐	☐	☐

Overall appraisal: Include ☐   Exclude ☐   Seek further info ☐
Comments (Including reason for exclusion)

_____
_____
_____

Explanation of textual evidence: expert opinion critical appraisal tool

How to cite: *McArthur A, Klugarova J, Yan H, Florescu S. Innovations in the systematic review of text and opinion. Int J Evid Based Healthc. 2015;13(3):188–195.*
   Answers: Yes, No, Unclear or Not/Applicable

1 Is the source of the opinion clearly identified?
To assess an opinion, it is important to locate its source. *Ask:*

- Are the authors clearly identified (Including their name, their role/ experience / qualifications)?

2 Does the source of the opinion have standing in the field of expertise?
Determining whether the author is informed or possesses knowledge about a specific subject is a key stage in assessing the credibility of the opinion. *Ask*:

- For health professionals or health researchers, what are their qualifications, current role and other indicators such as fellowships or licensures? Are any allegiances or affiliations with specific organisations or groups known?
- For patients/consumers/advocates, what are their experiences and role?

3 Are the interests of the relevant population the central focus of the opinion?
The expert opinion should focus on improving outcomes and it is important to determine that the opinion has such a focus. *Ask:*

- Does the paper take a position that advantages a profession or a specific institution or body; or financial or political objectives, rather than patients, clients, communities or health gain?

4 Does the opinion demonstrate a logically defended argument to support the conclusions drawn?

An opinion without a logical argument behind it is difficult to accept as a legitimate guide for practice/action. It is therefore important to look at the degree to which a logical argument to defend the conclusions drawn in the opinion is evident. *Ask:*

- Does the opinion 'make sense' and demonstrate an attempt to justify the stance it takes?
- Is the opinion the result of an analytical process drawing on experience or the literature?
- Does the argument comply with Toulmin's model for argumentation?

5 Is there reference to the extant literature?

It is important to determine whether or not the opinion expressed comes from a position of awareness of extant evidence. *Ask:*

- What extant literature does the author present to support the arguments?

6 Is any incongruence with the literature/sources logically defended?

Is there any reference provided in the text to ascertain if the opinion expressed has wider support?
Ask:

- Has the author demonstrated awareness of alternate or dominant opinions in the literature?
- Have they provided an informed defence of their position as it relates to other or similar discourses?

Checklist for textual evidence: narrative

Introduction

JBI is an international research organisation based in the Faculty of Health and Medical Sciences at the University of Adelaide, South Australia. JBI develops and delivers unique evidence-based information, software, education and training designed to improve healthcare practice and health outcomes. With over 70 Collaborating Entities, servicing over 90 countries, JBI is a recognised global leader in evidence-based healthcare.

JBI systematic reviews

The core of evidence synthesis is the systematic review of literature of a particular intervention, condition or issue. The systematic review is essentially an analysis of the available literature (i.e., evidence) and a judgment of the effectiveness or otherwise of a practice, involving a series of complex steps. JBI takes a particular view on what counts as evidence and the methods utilised to synthesise those different types of evidence. In line with this broader view of evidence, JBI has developed theories, methodologies and rigorous processes for the critical appraisal and synthesis of these diverse forms of evidence in order to aid in clinical decision-making in healthcare. There now exists JBI guidance for conducting reviews of effectiveness research, qualitative research, prevalence/incidence, aetiology/risk, economic evaluations, textual evidence, diagnostic test accuracy, mixed-methods, umbrella reviews and scoping reviews. Further information regarding JBI systematic reviews can be found in the JBI Evidence Synthesis Manual.

JBI critical appraisal tools

All systematic reviews incorporate a process of critique or appraisal of the research evidence. The purpose of this appraisal is to assess the methodological quality of a study and to determine the extent to which a study has addressed the possibility of bias in its design, conduct and analysis. All papers selected for inclusion in the systematic review (i.e., those that meet the inclusion criteria described in the protocol) need to be subjected to rigorous appraisal by two critical appraisers. The results of this appraisal can then be used to inform synthesis and interpretation of the results of the study. JBI critical appraisal tools have been developed by the JBI and collaborators and approved by the JBI Scientific Committee following extensive peer review. Although designed for use in systematic reviews, JBI critical appraisal tools can also be used when creating Critically Appraised Topics (CAT), in journal clubs and as an educational tool.

JBI critical appraisal checklist for textual evidence: narrative

Reviewer _____ Date _____

Author _____ Year _____ Record Number _____

	Yes	No	Unclear	Not applicable
1 Is the generator of the narrative a credible or appropriate source?	☐	☐	☐	☐
2 Is the relationship between the text and its context explained? (where, when, who with, how)	☐	☐	☐	☐
3 Does the narrative present the events using a logical sequence so the reader or listener can understand how it unfolds?	☐	☐	☐	☐
4 Do you, as reader or listener of the narrative, arrive at similar conclusions to those drawn by the narrator?	☐	☐	☐	☐
5 Do the conclusions flow from the narrative account?	☐	☐	☐	☐
6 Do you consider this account to be a narrative?	☐	☐	☐	☐

Overall appraisal: Include ☐   Exclude ☐   Seek further info ☐
Comments (Including reason for exclusion)

_____
_____
_____

Explanation of textual evidence: narrative critical appraisal tool

How to cite: *McArthur A, Klugarova J, Yan H, Florescu S. Innovations in the systematic review of text and opinion. Int J Evid Based Healthc. 2015;13(3):188–195.*
   Answers: Yes, No, Unclear or Not/Applicable

1 Is the generator of the narrative a credible or appropriate source?
   It is important to establish the legitimacy of the narrator as part of assessing the degree to which the narrative is authentic. *Ask:*

   • Is this a first-hand account of an event?
   • Do you sense that the author is both a credible and appropriate narrator?

2 Is the relationship between the text and its context explained?
   Narrative always describes an event that occurs within a specific time and space; within a context. The relationship between the characters and the place in which the event occurs needs to be described. *Ask:*

   • Where does the event take place?
   • Who does it involve?
   • What occurs?

3 **Does the narrative present the events using a logical sequence so the reader or listener can understand how it unfolds?**

A narrative seeks to convince a reader; this, in assessing this narrative, the reviewer should 'follow' the narrative and its meanings. **Ask:**

- Can I 'imagine' the event, the characters involved and what happened?
- Does the 'story' or the account flow in a logical way?

4 **Do you, as reader or listener of the narrative, arrive at similar conclusions to those drawn by the narrator?**

Again, note the purpose of narrative to persuade or convince. **Ask:**

- Are the conclusions drawn from the description of the event?
- Are any seemingly causal relationships explained?
- Do you draw similar conclusion from the narrative as the narrator?

5 **Do the conclusions flow from the narrative account?**

Again, note the purpose of narrative to persuade or convince. **Ask:**

- Are the conclusions drawn from the description of the event?

6 **Do you consider this account to be a narrative?**

In appraising the authenticity of the narrative, can you differentiate between the emotional persuasiveness of the 'story' with the objective accuracy of the narrative? **Ask:**

- What is the degree of narrativity in this piece?

Checklist for textual evidence: policy

Introduction

JBI is an international research organisation based in the Faculty of Health and Medical Sciences at the University of Adelaide, South Australia. JBI develops and delivers unique evidence-based information, software, education and training designed to improve healthcare practice and health outcomes. With over 70 Collaborating Entities, servicing over 90 countries, JBI is a recognised global leader in evidence-based healthcare.

JBI systematic reviews

The core of evidence synthesis is the systematic review of literature of a particular intervention, condition or issue. The systematic review is essentially an analysis of the available literature (i.e., evidence) and a judgment of the effectiveness or otherwise of a practice, involving a series of complex steps. JBI takes a particular view on what counts as evidence and the methods utilised to synthesise those different types of evidence. In line with this broader view of evidence, JBI has developed theories, methodologies and rigorous processes for the critical appraisal and synthesis of these diverse forms of evidence in order to aid in clinical decision-making in healthcare. There now exists JBI guidance for conducting reviews of effectiveness research, qualitative research, prevalence/incidence, aetiology/risk, economic evaluations, textual evidence, diagnostic test accuracy, mixed-methods, umbrella reviews and scoping reviews. Further information regarding JBI systematic reviews can be found in the JBI Evidence Synthesis Manual.

JBI critical appraisal tools

All systematic reviews incorporate a process of critique or appraisal of the research evidence. The purpose of this appraisal is to assess the methodological quality of a study and to determine the extent to which a study has addressed the possibility of bias in its design, conduct and analysis. All papers selected for inclusion in the systematic review (i.e., those that meet the inclusion criteria described in the protocol) need to be subjected to rigorous appraisal by two critical appraisers. The results of this appraisal can then be used to inform synthesis and interpretation of the results of the study. JBI critical appraisal tools have been developed by the JBI and collaborators and approved by the JBI Scientific Committee following extensive peer review. Although designed for use in systematic reviews, JBI critical appraisal tools can also be used when creating Critically Appraised Topics (CAT), in journal clubs and as an educational tool.

JBI critical appraisal checklist for textual evidence: policy/consensus guidelines

Reviewer _____ Date _____

Author _____ Year _____ Record Number _____

	Yes	*No*	*Unclear*	*Not applicable*
1 Are the developers of the policy/ consensus guideline (and any allegiances/affiliations) clearly identified?	□	□	□	□
2 Do the developers of the policy/ consensus guideline have standing in the field of expertise?	□	□	□	□
3 Are appropriate stakeholders involved in developing the policy/guideline and do the conclusions drawn represent the views of their intended users?	□	□	□	□
4 Are biases due to competing interests acknowledged and responded to?	□	□	□	□
5 Are the processes of gathering and summarising the evidence described?	□	□	□	□
6 Is any incongruence with the extant literature/evidence logically defended?	□	□	□	□
7 Are the methods used to develop recommendations described?	□	□	□	□

Overall appraisal: Include □   Exclude □   Seek further info □
Comments (Including reason for exclusion)

_____
_____
_____

Explanation of textual evidence: policy/consensus guidelines critical appraisal tool

How to cite: *McArthur A, Klugarova J, Yan H, Florescu S. Innovations in the systematic review of text and opinion. Int J Evid Based Healthc. 2015;13(3):188–195.*
  Answers: Yes, No, Unclear or Not/Applicable

1 Are the developers of the policy/consensus guideline (and any allegiances/affiliations) clearly identified?
  To assess a policy or guideline that seeks to direct action, it is important to be aware of who was involved in its development. *Ask:*

- Are the authors clearly identified (Including their name, their role/experience/ qualifications?).
- Are any allegiances or affiliations with specific organisations or groups known?

2 **Do the developers of the policy/consensus guideline have standing in the field of expertise?**
Determining whether the developers are informed or possess knowledge about a specific subject is a key stage in assessing the credibility of a policy or guideline. *Ask*:

- For health professionals or health researchers, what are their qualifications, current role and other indicators such as fellowships or licensures? (Reviewers may wish to follow up the standing of the source by consulting with experts in the field of expertise; checking accreditation rolls; or contacting the source for further information.)
- For patients/consumers/advocates, what are their experiences and role?

3 **Are appropriate stakeholders involved in developing the policy/consensus guideline and do the conclusions drawn represent the views of their intended users?**
Guideline and policy development requires involvement of (or at least consultation with) both health care providers who will be expected to implement them and the receivers of healthcare (patients/clients/consumers). *Ask:*

- Who are the central stakeholders that might be impacted by this policy/guideline?
- Are these stakeholders either part of the development group; or is there evidence that they have been consulted?

4 **Are biases due to competing interests acknowledged and responded to?**
All policy/guideline development groups are likely to include competing interests and to be subject to a range of biases. The quality of the development process is improved if competing interests and potential biases are identified and addressed. *Ask:*

- Are potential competing interests identified in the policy/guideline document?
- Are potential biases identified in the policy/guideline document?
- Are any strategies to acknowledge and address competing interests and biases presented in the policy/guideline document?

5 **Are the processes of gathering and summarising the evidence described?**
Some policy/guideline developers search for and use published evidence reviews (systematic reviews, etc.), published and unpublished papers; and local clinical and activity data. Others commission a full evidence review. For our purpose, it is important to assess the quality of gathering and summarising data. *Ask:*

- Are the processes involved in gathering and analysing extant evidence detailed?
- Are the approaches taken rigorous?

6 **Is any incongruence with the extant literature/evidence logically defended?**
Whilst policy/guideline developers may search for and refer to synthesised evidence and because of possible competing interests and local biases, the external evidence may not concur with the conclusions or recommendations embodied in the resulting policy or guideline. *Ask:*

- Is there any incongruence between the conclusions/recommendations and the extant literature?
- If there is, is this acknowledged in the paper/document?
- Is there a logical defence of any position taken that is in conflict with the extant literature?

7   **Are the methods used to develop recommendations described?**
Policy and guideline developers usually spend a great deal of time and exert much effort on developing final conclusions or recommendations and seek to balance the evidence with the expertise of the development group and the views of other stakeholders (frequently seeking a consensus view). Thus, a description of how recommendations or conclusions are developed is of importance. *Ask:*

- Is the process of developing recommendations or conclusions documented?
- Do these processes suggest that a balance between opinion and evidence was sought?

4 Impact of funding agencies on the production of research

Abstract

The impact that funding agencies and their criteria for funding research has on scholars and their creation of knowledge are explored in this chapter. It begins by explaining strategies that can make the process of applying for grant funding less cumbersome and bureaucratic for both grant agencies and prospective applicants. Strategies that funding agencies can use to improve the quality of research are then explained. These strategies include improving the evaluation of knowledge translation in research proposals, addressing the gender bias in research, preventing funding agencies from influencing the creation of research, and making research reproducible. The purpose of this chapter is to explain how funding agencies can improve the quality of research.

Keywords: Funding Agencies; Incumbency Advantage; Reproducibility; Translational Research; Two-Stage Application Process

Key points

- Agencies that fund research can be inundated with funding applications that can undermine a comprehensive assessment of the application. A two-stage application process can reduce this volume of paperwork and the workload placed on peer reviewers.
- In comparison to inexperienced applicants, those with experience typically obtain funding grants. To reduce this 'incumbency advantage' a proportion of grant funding should only be allocated to graduates who have obtained their PhD in the last five years. Alternatively, a lottery for awarding grants, grant coaching programmes, and good quality feedback on failed applications can also assist inexperienced applicants obtain funding opportunities.
- Typically, funding agencies do not finance studies that are translational or reproducible. To prevent such research being created grant agencies should mandate that funding proposals contain details about how the study's results will be both translational and reproducible.

4.1 Improving the grant application process

4.1.1 Implementing a two-stage application process

Often researchers use considerable resources to create grant applications that will most likely never be funded. This point was explored by Herbert and colleagues (2013) who examined the outcome of 3,727 grant applications that were submitted to the Australian Government's

DOI: 10.4324/9781003510376-4

National Health and Medical Research Council (NHMRC) in March 2012. They reported that out of the 3,727 grant applications submitted, 3,570 were reviewed and out of those 731 (21%) were funded. They also reported that it took academics on average of 38 days to create a grant application and 28 days to implement the modifications requested by the NHMRC. Based on these statistics, Herbert and colleagues estimated that 550 years had been used to prepare the 3,727 grant applications. Due to these findings, Herbert and colleagues (2013, p. 1) concluded that:

> Considerable time is spent preparing NHMRC Project Grant proposals. As success rates are historically 20–25%, much of this time has no immediate benefit to either the researcher or society, and there are large opportunity costs in lost research output. The application process could be shortened so that only information relevant for peer review, not administration, is collected. This would have little impact on the quality of peer review and the time saved could be reinvested into research.

A two-stage application process can reduce the paperwork and resources academics use to apply for a grant, including NHMRC grants. The NHMRC, for example, uses a one-stage grant application process. Typically, each application to the NHMRC is between 80 and 120 pages in length and reviewers are typically expected to examine between 50 and 100 applications. Due to this workload 'it is optimistic to expect accurate judgements in this sea of excessive information' (Herbert et al., 2013, p. 5). To reduce this burden the NHMRC can implement a two-stage application process, whereby for the first stage the researcher submits a brief proposal and if the proposal passes then they submit for the second stage a full grant application. A two-stage application process can reduce the paperwork that researchers produce because if they do not pass the first stage then they do not need to create a full grant application. To further reduce this paperwork the NHMRC can impose maximum word limits for grant applications submitted for both the first and second stage. For example, for the first stage the word count could be 5,000 words and for the second stage it could be 15,000 words. Failure to adhere to these mandatory guidelines would result in an automatic rejection of the proposal. Also, with such word limits peer reviewers would examine succinct proposals which would reduce their burden and the potential of them making evaluation errors.

A two-stage application process has been implemented by some funding agencies and has yielded promising results. In 2015, the United Kingdom's *National Institute of Health Researchers* replaced its two-stage application process with a one-stage application process for the *Research for Patient Benefit Programme* grant. This change gave Morgan and colleagues (2020) the opportunity to compare the efficiency of both their one- and two-stage application processes. They reported that it took a mean of 274 days to review proposals submitted to the one-stage application process and 348 days for the two-stage application process. For the one-stage application process a mean of 423 peer reviewers and 102 lay reviewers were used to review grant applications and for the two-stage application process this mean was 208 peer reviewers and 50 lay reviewers per funding round. The overall average cost for a one-stage application process was £148,908 while for the two-stage application process this average cost was £105,342, which was an average difference of £43,566 per funding round. Based on these statistics, Morgan and colleagues (2020, p. 2) concluded:

> that a two-stage application process increases the number of applications submitted to a funding round, is less burdensome and more efficient for all those involved with the process, is cost effective and has a small increase in peer reviewer scores.

4.1.2   *Overcoming the 'incumbency advantage'*

Compared to inexperienced researchers, those with a history of obtaining funding grants are more likely to receive funding grants in the future (Ballabeni et al., 2016). Ballabeni and colleagues (2016) have provided three reasons to explain this 'incumbency advantage'. First, regardless of the merit or novelty of their research, inexperienced researchers struggle to create a competitive proposal because they often do not have access to a large amount of preliminary data to support their application. In contrast, established researchers typically have access to large datasets that they can use to argue that their proposals should be funded. Second, newer researchers lack the experience, knowledge, and skills to interpret lengthy grant application documents that contain complex guidelines and criteria. In contrast, experienced researchers can competently navigate such elaborate documents. Third, agencies often take a conservative approach and fund mainstream ideas proposed by established researchers that have a proven track record of research output instead of unorthodox ideas proposed by newly minted researchers (Ballabeni et al., 2016).

Funding agencies can adopt two policies to reduce the detrimental consequences that the 'incumbency advantage' can have on inexperienced researchers. First, they can create funding opportunities that are only available to researchers with less than five years post-PhD research experience. Second, grant applications submitted must provide details about the tasks that early career researchers will perform in the project and the research team. To ensure that their inclusion is not tokenistic as a condition of receiving funds grant proposals must contain reassurances that early career researchers will perform tasks that more experienced researchers typically perform, such as meeting key stakeholders and dissemination of the study's results at conferences (Ballabeni et al., 2016). Such guarantees can ensure that early career researchers receive adequate opportunities and experience that can help them progress their careers to leadership roles.

A lottery for allocating grant funding can improve the chances of inexperienced researchers receiving grants (De Peufer & Conix, 2021; Liu et al., 2020). The *Health Research Council of New Zealand* has implemented a lottery system to allocate funding to researchers who applied for the Explorer Grant scheme. To test the effectiveness of this scheme, Liu and colleagues (2020) asked 126 applicants who had applied for funding from the Explorer Grant scheme from 2013 to 2019 about their views of using a lottery to allocate research funding. They found that in response to the question 'Do you think the randomisation process is an acceptable method of allocating Explorer Grant funds?' most answered 'Yes' (i.e., Yes = 63%, Unsure = 12%, and No = 25%). Additionally, in response to the question 'Do you think a randomisation process would be an acceptable method for the allocation of funding for other grant types?' most also answered 'Yes' (i.e., Yes = 40%, Unsure = 24%, and No = 37%). However, most answered that they would not write their application differently if they knew that funding could be randomly allocated (i.e., Yes = 25%, Unsure = 6%, and No = 69%). Finally, most would not change the amount of time that they would use preparing their application if they knew that there was a random allocation of funding (i.e., Less time = 16%, No difference = 75%, More time = 5%, and Unsure = 5%). Based on these responses, researchers appear comfortable with a random allocation of funding should their application progress to the final stage.

Education can give some inexperienced scholars the opportunity to improve their grant writing skills, which will help them overcome the 'incumbency advantage'. To give inexperienced researchers from marginalised groups a better chance of winning a research grant the *National Research Mentoring Network* developed a *Grant Writing Coaching Program* (GCP) (see Table 4.1). Weber-Main and colleagues (2020) examined if the GCP improved the likelihood of researchers from marginalised backgrounds obtaining research grants. The results of their study were based on 545 investigators (i.e., 67% females and 61% under-represented racial/ethnic minorities) from 187 different institutions who participated in the GCP. Within this

Table 4.1 Unique features of the NRMN Grant Writing Coaching Programme models

	Model	Target audience of investigators	Coaching duration	Cohort size	Typical coach-to-participant ratio	Unique approaches
Less experienced investigators	STAR	No or minimal grant writing experience Majority from minority serving institutions	12 month	10	1:2	In addition to grant writing coaching, cohort engages in professional development sessions on essential skills .for research careers in academia. Blend of virtual and in person meetings. ending with a mock study section.
	CAN	No or minimal grant writing experience All from institutions. In Big Ten Academic Alliance	6 month	40	1:4	Launch of coaching groups is preceded by professional development sessions for academic career preparation. After large-group kick-off, coaching groups meet in person on their home campuses.
	SETH	No or minimal grant writing experience Majority from RCMIs	6 month	35	1:4	Coaching groups are augmented by the Health Equity Learning Collaboratory a structured virtual online community that provides trainees direct access to NH. Resources and peer networks.
	GUMSHOE	Minimal to moderate grant writing experience Target population defined for each cohort	6 month	25	1:1	Cohorts have a distinct population focus in health disparities research. Extensive engagement with NH grant programme officials. One-on-one coaching occurs after kick-off. Dyads work through all proposal sections, conducting with a mock study section.
More experienced investigators	NU	Minimal to moderate grant writing experience Preferably ready to write NIH K or R proposal (new/revised)	4–6 month	25	1:4	Coaches provide real-time iterative feedback to small group of trainees. Strong emphasis on rhetorical patterns common to many NIH-style proposals. Groups work on Aims, Significance, Innovation, and Career Development sections of proposals.
	P3	Moderate grant writing experience All ready to write NIH K or R proposal (new/revised)	5 month	10	1:2	Multiple coaches work with a single cohort of trainees, all developing NIH proposals, highly structured to enhance accountability and writing progress. Group works through all proposal sections, concluding with a mock study section.

Source: Weber-Main et al. (2020, p. 6).

sample 324 (59%) participants submitted at least one grant application and of those 134 (41%) received funding. Based on these outcomes, Weber-Main and colleagues concluded that the GCP can be an effective programme that can assist researchers from marginalised groups obtain grant funding.

4.1.3   Multiple application opportunities to reduce application burden and stress

In Australia, the NHMRC only accepts applications for research funding once a year. This annual deadline has contributed to some applicants feeling anxiety, stress, and depression (Herbert et al., 2014). This point was confirmed by Herbert and colleagues (2014, p. 8), who stated:

> The process of preparing grant proposals for a single annual deadline is stressful, time consuming and conflicts with family responsibilities. The timing of the funding cycle could be shifted to minimise applicant burden, give Australian researchers more time to work on actual research and to be with their families.

To reduce the prospects of this stress developing funding agencies can create multiple opportunities per annum for applicants to submit their grant proposals (Herbert et al., 2014). In conjunction with this recommendation, funding agencies can impose word limits for the proposal. Such limits will reduce the stress associated with creating lengthy applications and the paperwork burden that reviewers encounter (Herbert et al., 2013).

4.1.4   Improving the quality of feedback that unsuccessful applicants receive

Since they are examining a considerable number of lengthy and complex grant applications, reviewers often cannot provide clear and comprehensive feedback to applicants. Herbert and colleagues (2013) explained that reviewers often assess between 50 and 100 NHMRC applications per funding round. Due to this workload reviewers often cannot give feedback that applicants would consider to be of use for grant applications in the future. The NHMRC is not the only funding agency that has this issue. Gallo and colleagues (2021) discovered that there were no studies about if grant applicants thought that the feedback that they had received about their grant applications by reviewers was useful for future applications. To address this lack of research, with the assistance of scientists, scientific review administrators, research funders, and insights from the literature they created a survey. They distributed their survey to 13,091 potential respondents, of which 74% had applied in the previous three years for research funding. Their survey was emailed to potential respondents in September 2016, and it remained open for two months. After a reminder, 1,231 (9.4%) of all potential respondents completed the entire online survey. They reported that less than 40% of applicants they had surveyed thought that the criticisms that they had received about their grant application were helpful in improving their grant applications in the future.

A lack of clear and comprehensive feedback can undermine an inexperienced applicant's prospect of obtaining a funding grant in the future. With this hurdle those who have the skills and experience needed to obtain a research grant will continue to have an advantage with respect to receiving grants. To equalise the chances for both experienced and inexperienced applicants receiving research grants, funding agencies should ensure that unsuccessful applicants receive accurate and detailed explanations as to why they did not receive a grant. Armed with this information they can refine their grant writing skills and improve their chances of obtaining a research grant in the future. To maximise the prospect of this outcome occurring, funding

agencies can educate reviewers in what constitutes valuable criticism for unsuccessful applicants, such education can occur through workshops and guides that contain examples of both
valuable and worthless comments for unsuccessful applicants. Additionally, to determine if
reviewers have learnt how to give unsuccessful applicants useful feedback funding agencies can
distribute Gallo et al.'s survey to unsuccessful applicants (see Appendix 4.1).

4.2 How funding agencies can improve the quality of research

4.2.1 Improving the evaluation of knowledge translation in research proposals

Historically, a grant proposal's scientific merits were used to determine if it would be funded. However, contemporary criteria for deciding which grant proposals should receive funding are influenced by other considerations, including the potential for the study's results to be commercialised
or if it will influence the development of evidence-informed policy. Furthermore, some agencies are
starting to require that grant proposals contain a knowledge translation plan, otherwise known as a
dissemination and implementation plan. The purpose of such plans is to improve the prospect of the
knowledge gained from the study being disseminated to a broader audience than just academics.
Such criteria should be applauded since it seeks to improve the impact and relevance that research
can have on the community that funds its creation. However, reviewers may not be skilled to assess
a study's potential knowledge translation. Similarly, agencies may not have the institutional knowledge, capacity, or experience to also review this potential (Scarrow et al., 2017).

Five recommendations have been proposed to improve the ability of reviewers and agencies
to examine a proposal's knowledge translation potential. First, grant agencies should provide
reviewers with resources, such as the *Guide for Assessing Health Research Knowledge Translation*, that they can use to help them assess the potential knowledge translation of a study
(Scarrow et al., 2017). Second, grant agencies should arrange meetings between previous and
current reviewers so that they can exchange information about how to assess a grant proposal's knowledge translation aspects. Third, grant agencies can incorporate specific criteria about
knowledge translation into their guidelines and funding applications (Ruppertsberg et al., 2014).
If authors adhere to these guidelines the task of reviewers evaluating a research proposal's
knowledge translation potential should be easier and more efficient. Fourth, grant agencies can
employ knowledge translation experts to help reviewers evaluate the knowledge translation
potential in a research proposal (Scarrow et al., 2017). Fifth, researchers should incorporate into
their proposals strategies to reduce any barriers to translational research. Senecal and colleagues
(2021) have illustrated these barriers (see Figure 4.1).

4.2.2 Addressing sex and gender bias in research

In some fields there is a lack of specificity about the participants' sex and gender. Day and colleagues (2019), for example, examined the sex and gender of participants in 155 studies about
diabetes. They reported that while most articles ($n=151/155$, 97.4%) contained descriptions
of the participants' sex and gender, only ten (6.5%) contained explanations about how the participants' sex and gender impacted the results. To ensure that the participants' sex and gender
is adequately explained, as a condition of receiving funds grant agencies should mandate that
researchers capture and report this information. For example, the *Canadian Institutes of Health
Research* has implemented a series of policies intended to ensure that researchers describe the
participants' sex and gender in their grant proposals. These policies have improved the reporting
of this information (see Figure 4.2).

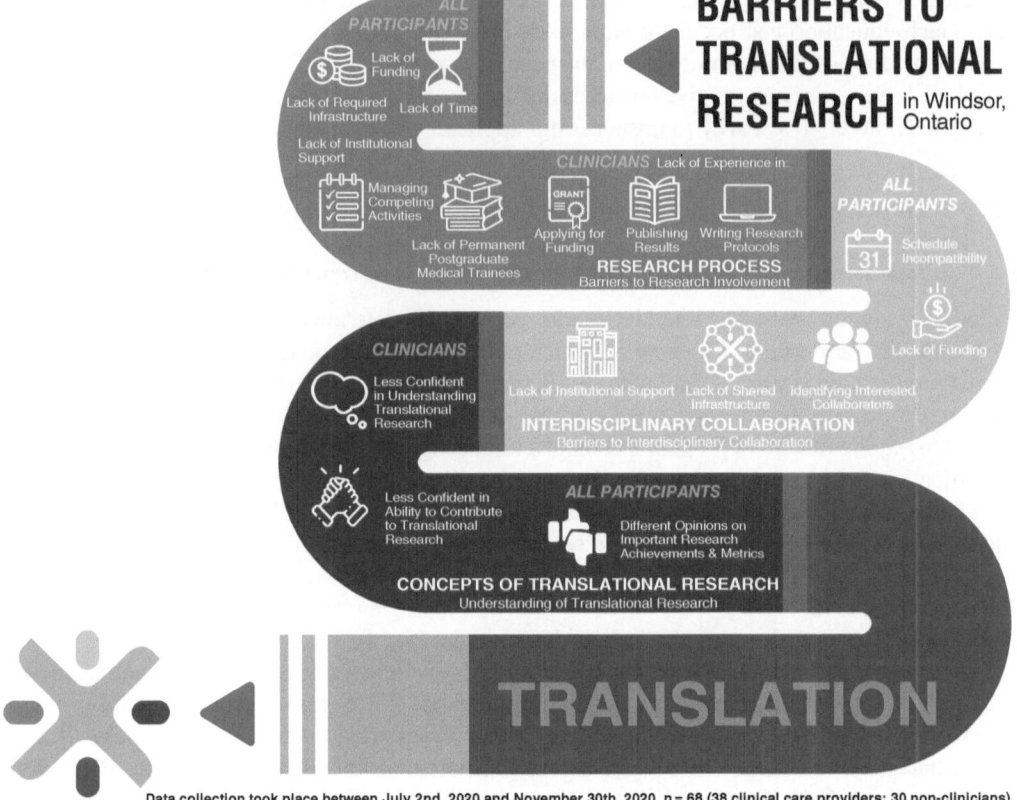

Figure 4.1 Barriers to translational research in Windsor, Ontario

Note: Summary of barriers to translational research important to clinicians and non-clinicians surveyed in our study (*n*=68). Barriers are sorted into three thematic barriers to translational research broadly identified as 'Research Process', 'Interdisciplinary Collaboration,' and 'Concepts of Translational Research'.

Source: Senecal et al. (2021, p. 8).

Aside from grant agencies, some researchers need some guidance about how to include the participants' sex and gender in their grant proposals. To assist them Hankivsky and colleagues (2018) have created a checklist that researchers can use to ensure that they have incorporated into their grant proposals techniques for recording and explaining this data. Furthermore, Mason (2020) has provided 11 strategies that researchers can use to improve their integration of sex and gender in grant proposals about health research (see Table 4.2).

It is hoped that by using Mason's strategies and the checklist provided by Hankivsky and colleagues researchers will be able to improve their reporting of the participant's sex and gender in both their grant proposals and studies.

4.2.3   *Preventing inappropriate influence by funding agencies*

Sometimes funding agencies coerce researchers to change their study and/or results so that they adhere to the funder's expectations. McCrabb and colleagues (2021) reported that 18% of the respondents in their study revealed that they had experienced at least one of seven predefined

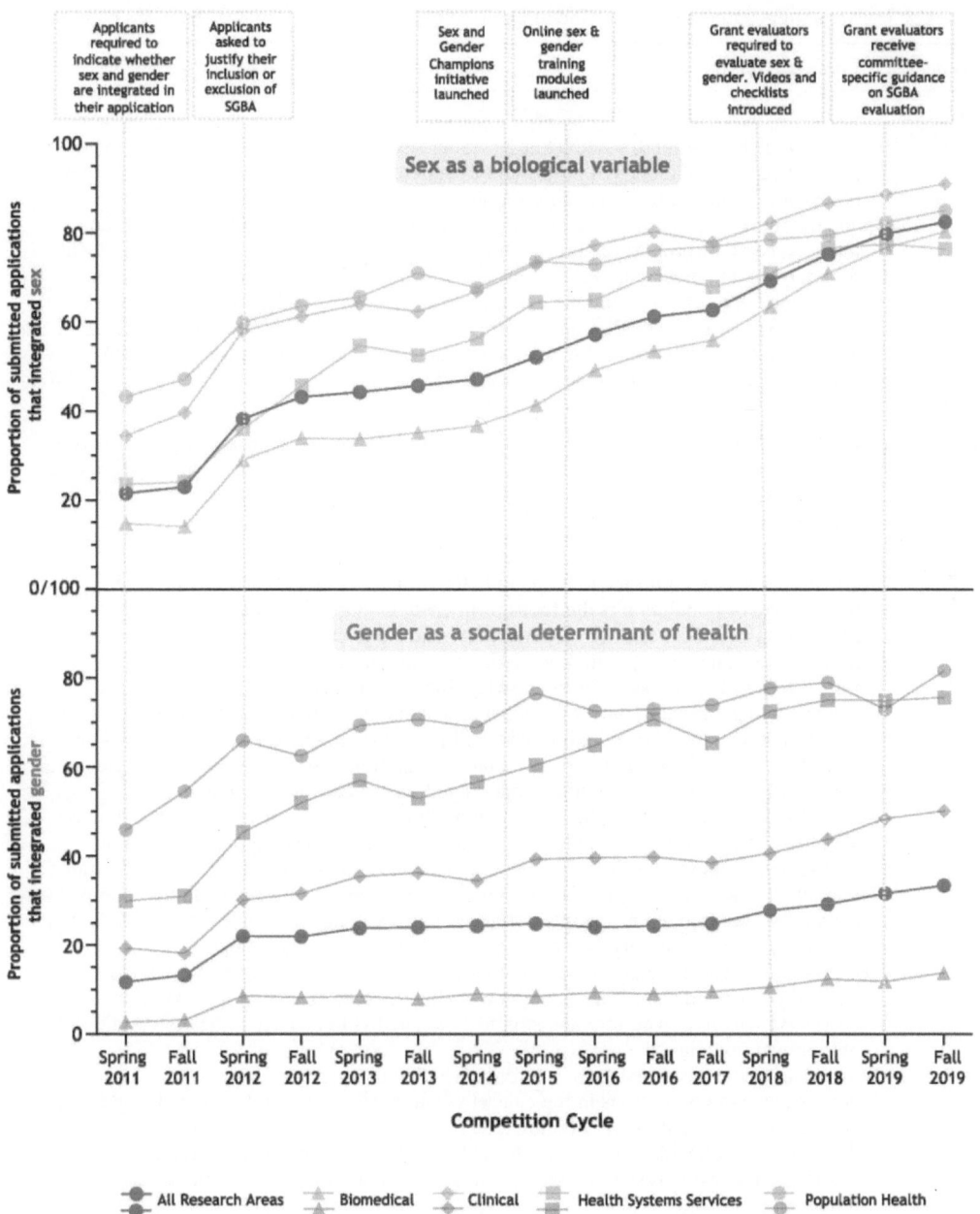

Figure 4.2 Integration of sex and gender in grant proposals submitted to Canadian Institutes of Health Research's largest investigator-initiated competition from 2011 to 2019

Note: Data represent the proportion of applications submitted to each investigator-initiated competition cycle between 2011 and 2019 that accounted for sex and gender, disaggregated by research area. Dotted overlays highlight key SGBA interventions implemented by CIHR at several time points throughout the ten-year period.

Source: Haverfield and Tannenbaum (2021, p. 6).

Table 4.2 Mason's suggestions to improve the integration of sex and gender in health research proposals

Number	Recommendation
Background and literature review	
1	Include in the background, what is known about sex and/or gender issues related to prevalence, presentation, symptoms and treatments for the issue or condition of interest. Note gaps in knowledge when appropriate.
2	In the literature review, note whether past studies integrated a sex and gender lens and included equal or proportional representation of males and females, disaggregated data for analysis, and presented findings by sex, irrespective of whether or not differences were found. Comment on the inclusion or absence of sex/gender, gender identity or other relevant factors in the demographic form, recruitment strategies, results or discussion. The concluding sentence of this section can note sex and gender-related gaps and strengthen the rationale for the proposed study.
Goal and objectives	
3	Clearly state who is likely to benefit from the study. Will men, women and gender-diverse individuals equally benefit? There is tremendous heterogeneity among men and among women, is the study equally important to all populations? Secondary objectives could explore some of these questions.
Methods	
4	Form an Advisory Committee to help guide the project, especially when a specific health issue or population is the focus of interest. Community members and other relevant stakeholders can help ensure study goals, objectives and methods are relevant, appropriate, and can help with knowledge translation activities.
5	Describe the study population by acknowledging sex and gender.
6	Consider stating, 'equal numbers of men and women' or 'both men and women' or 'men, women and gender diverse individuals' will be recruited, instead of 'patients will be recruited'. Sample size should be calculated to, at minimum, support sex disaggregated analysis. Ideally, analysis should be conducted separately for men and women.
7	Consider the language used in the demographic form. If sex and gender are relevant, ask about sex assigned at birth (male, female, and intersex) and current gender identity (a drop-down menu may be helpful). Ensure language is inclusive of those whose gender identity is fluid or non-binary, and provide culturally and ethnically suitable options. For example, some Indigenous people describe their sexual, gender and/or spiritual identity as 'two-spirited'. Avoid using the word, "other" as an open-ended option.
8	Design recruitment strategies to accommodate those with caregiving responsibilities (should you provide childcare?), income challenges (can you pay for parking or travel costs). Be sensitive to culture and stigma associated with identity or health condition in recruitment materials.
9	When using pre-existing datasets, reference whether disaggregation by sex and gender is possible. Note limitations in the dataset when this is not possible. Consider adding a qualitative component when existing data has no reference to gender and other social determinants of health. If this is not feasible, suggest this may be explored in future studies. Indicate that results will report on sex and gender even when no differences are found.
Dissemination and knowledge translation	
10	Note when opportunities to publish and present at conferences presentation will be made available to all investigators. Describe the ways in which results will be tailored to men, women and relevant sub-populations. Provide examples to enhance the knowledge translation section of the proposal.
Team description	
11	Include both men and women on the study team. Ensure that both men and women are identified in the same way, that is, with or without their titles such as Dr or Professor. Consider whether the descriptions of all team members are roughly comparable in length style and substance.

Source: Mason, 2020.

coercion tactics. The most reported coercion tactic was "Funder expressed reluctance for publication because they considered the results 'unfavourable'" (see Table 4.3).

Multiple stakeholders need to protect researchers against the coercive influence that some funding agencies may exert. Government can draft and enforce legislation and regulations that protect academic freedom. For example, removing clauses that require a funding agency's approval of results prior to publication could be legislated. It should become an expectation that research institutions do not accept funding from donors that want the ability to interfere in the production and/or reporting of the study's results. It should also become standard practice that academic journals mandate that authors declare all their sources of funding and any potential conflicts of interest. Governments should fund agencies that oversee research integrity in research institutions and tertiary academic institutions (McCrabb et al., 2021). Within some jurisdictions such agencies exist. For example, in the United Kingdom, the *UK Research Integrity Office* provides support and guidance to the public, researchers, and organisations to improve the practice of academic, scientific, and medical research (UK Research Integrity Office, 2022). Finally, tertiary academic institutions should either establish or improve existing mechanisms for reporting instances of research misconduct or suppression (McCrabb et al., 2021). For example, universities can either employ or improve the abilities of research integrity officers to examine accusations about academic misconduct or suppression (Kalichman, 2020; Maisonneuve, 2019).

4.2.4   *Making research reproducible*

As will be elaborated in Chapter 7, for a study's results to be confirmed identical results need to be created after the same analysis has occurred with the same dataset. When this process does not occur, the study is deemed to be 'irreproducible', and its results will forever be doubted since they cannot be confirmed. Ensuring that research is reproducible might require modifications to any components of the research production process, such as researchers clearly articulating the processes that they used to examine the dataset and journals mandating that datasets be published along with the study. The focus of this chapter will be on the responsibilities that grant agencies have in ensuring that the research they fund is reproducible.

To improve the prospects of research being reproducible grant agencies should mandate that applicants incorporate into their research design factors that can ensure that their work is reproducible. Diong and colleagues (2021) measured the extent that eight health and medical funding programmes in Australia had in incentivising responsible research practices. They purposively selected these programmes because at the time they distributed the largest amount of funds in Australia. Each scheme was assessed against nine criteria about reproducibility and responsible research and reporting practices. These nine criteria were:

1 publicly register study protocols before starting data collection,
2 register analysis protocols before starting data analysis,
3 make study data openly available,
4 make analysis code openly available,
5 make research materials openly available,
6 discourage use of publication metrics,
7 conduct quality research (e.g., adhere to reporting guidelines),
8 collaborate with a statistician, and
9 adhere to other responsible research practices.

Table 4.3 Researcher reports of funder efforts to funder efforts to suppress

Funder type	Never	Once or more often*					
		Industry	Other government agency	Philanthropic	Dedicated research agency	Multiple	Unknown
Funder expressed reluctance for publication because they considered the results 'unfavourable'	89	0	6	0	2	1	0
Funder delayed reporting of findings until a more favourable time (e.g., following elections)	93	0	2	0	0	2	1
Funder asked researcher to alter conclusions to better align with funder interests	91	0	3	0	1	2	0
Funder asked researcher to not report findings they considered unfavourable	95	0	2	0	1	0	0
Funder discouraged researcher from presenting results to certain groups or organisations that may have an interest in the intervention	95	0	1	1	1	1	0
Funder attempted to discredit members of the research team or other staff involved in the conduct of the study	94	0	1	0	2	1	0
Funder demanded changes to study methods or analysis likely to produce findings that aligned with funder interests (e.g., emphasis on the 'statistical significance' of a result)	94	0	1	0	1	0	0

Source: McCrabb et al. (2021, p. 6).

Note: *Six respondents did not answer any of these questions, while the number of respondents answering each question ranged from 96 to 98.

Table 4.4 Diong et al.'s criteria for examining the reproducibility mandates of fundings agencies

Question	Explanation
1 Do instructions incentivise publicly registering study protocols before starting data collection?	Instructions must state that study protocols must be publicly registered with a date and time-stamp before starting data collection.
2 Do instructions incentivise registering analysis protocols before starting data analysis?	Instructions must state that analysis protocols must be registered with a date and time-stamp before starting data collection.
3 Do instructions incentivise making study data openly available to the research community?	Instructions must state that data must be made openly or publicly available.
4 Do instructions incentivise making analysis code openly available to the research community?	Instructions must state that computer code used to analyse the data must be made openly or publicly available.
5 Do instructions incentivise making research materials openly available to the research community?	Instructions must state that materials used to conduct the research must be made openly or publicly available. Examples include but are not limited to supplemental appendices, questionnaires, survey instruments, scoring rubrics, visual stimuli, and scripts used by research personnel.
6 Do instructions incentivise the discouragement of publication metrics?	Instructions must state that publication metrics (e.g., impact factor and H-index) should not be used.
7 Do instructions incentivise research quality (e.g., adherence to reporting guidelines)?	Instructions must state at least one mechanism to promote research quality that was not part of former criteria. Examples include but are not limited to adhering to reporting guidelines, adhering to ethical standards, avoiding or acknowledging biases, prioritising robust methodology, broadening research dissemination, and maximising the value and impact of all research output.
8 Do instructions incentivise collaboration with a statistician?	Instructions must state that applicants consult a statistician for complex quantitative analyses.
9 Do instructions incentivise any other responsible research practices?	Instructions must state any at least one other responsible research practice that was not covered in Question 7. Examples include but are not limited to research training, and declaring conflicts of interest.

Source: Diong et al. (2021).

Using these nine criteria as a guide, Diong and colleagues examined each scheme with a series of questions (see Table 4.4). Each question was answered using one of three responses, which were 'Instructed', 'Encouraged', and 'No mention'. They reported that no grant scheme recommended that study protocols should be publicly registered before the start of data collection (i.e., Question 1). Additionally, no grant scheme mandated that the analysis protocols should be registered before the start of the data analysis (i.e., Question 2), make analysis codes openly available (i.e., Question 4), or collaborate with a statistician (i.e., Question 8). In contrast, most grant schemes encouraged scholars to make their dataset available (i.e., Question 3) and make research materials available (i.e., Question 5). Finally, most schemes either encouraged or instructed researchers to adhere to reporting guidelines (i.e., Question 7) (see Table 4.5).

Based on the results collected most funding agencies in Australia did not require applicants to engage in research practices that foster the reproducibility of results. To ensure that results can be verified funding agencies should require that all applicants adhere to the nine

Table 4.5 Incentives for responsible research practices in each funding scheme

Do funding instructions incentivise applicants to:	ARC discovery projects	ARC DECRA	NHMRC project grants	NHMRC CDF	NHMRC CTCS	DART	NBCF	Bupa Health Foundation
1 Publicly register study protocols before starting data collection	No mention	No mention	No mention	No mention	Instructed	No mention	No mention	No mention
2 Register analysis protocols before starting data analysis	No mention	No mention	No mention	No mention	No mention	No mention	No mention	No mention
3 Make study data openly available	Encouraged	Encouraged	Encouraged	Encouraged	Encouraged	No mention	No mention	No mention
4 Make analysis code openly available	No mention	No mention	No mention	No mention	No mention	No mention	No mention	No mention
5 Make research materials openly available	Encouraged	Encouraged	Encouraged	Encouraged	Encouraged	No mention	No mention	No mention
6 Discourage use of publication metrics	No mention	No mention	Instructed	Instructed	Instructed	No mention	No mention	No mention
7 Conduct quality research (e.g., adhere to reporting guidelines)	Instructed	Instructed	Encouraged	Encouraged	Encouraged	Encouraged	Encouraged	No mention
8 Collaborate with a statistician	No mention	No mention	No mention	No mention	No mention	No mention	No mention	No mention
9 Adhere to other responsible research practices	Encouraged	Encouraged	No mention	No mention	No mention	No mention	No mention	No mention

Source: Diong et al. (2021, p. 4).

Abbreviations: ARC, Australian Research Council; *CDF,* Career Development Fellowship; *CTCS,* Clinical Trials and Cohort Studies; *DART,* Diabetes Australia Research Trust General Grants and Millennium Awards; *DECRA,* Discovery Early Career Researcher Award; *NBCF,* National Breast Cancer Foundation Investigator Initiated Research Scheme; *NHMRC,* National Health and Medical Research Council.

principles about the responsible conduct of research that Diong and colleagues outlined. Failure to adhere to these principles will result in the funding of research in which results cannot be confirmed.

4.3　Conclusion

The purpose of this chapter was to explain the influence that funding agencies have on the creation of research and how their role can be changed to improve the allocation of research funding. It began with an explanation about some strategies that can improve the efficiency of both researchers applying for grant funding and agencies allocating such funding. The focus of this chapter then changed to explaining strategies that funding agencies can implement to improve the quality of research that is produced. These strategies include making as a condition of receiving funding the reporting of the participants' sex and gender in grant proposals and incorporating practices that enhance reproducibility. By implementing the suggestions in this chapter it is possible that the quality of research will improve, and funding will be more equitably distributed.

Additional readings

Bendiscioli, S. (2019). The troubles with peer review for allocating research funding: Funders need to experiment with versions of peer review and decision-making. *EMBO Reports, 20*(12), e49472. https://doi.org/10.15252/embr.201949472

Guyer, R. A., Schwarze, M. L., Gosain, A., Maggard-Gibbons, M., Keswani, S. G., & Goldstein, A. M. (2021). Top ten strategies to enhance grant-writing success. *Surgery, 170*(6), 1727–1731. https://doi.org/10.1016/j.surg.2021.06.039

Perry, J. J., Vaillancourt, C., Hohl, C. M., Thiruganasambandamoorthy, V., Morris, J., Emond, M., Lee, J., & Stiell, I. G. (2021). Optimizing collaborative relationships in emergency medicine research. *CJEM, 23*(3), 291–296. https://doi.org/10.1007/s43678-020-00080-w

Santana, C. (2022). Why citizen review might beat peer review at identifying pursuitworthy scientific research. *Studies in History and Philosophy of Science, 92*, 20–26. https://doi.org/10.1016/j.shpsa.2022.01.012

Viergever, R. F., & Hendriks, T. C. (2016). The 10 largest public and philanthropic funders of health research in the world: What they fund and how they distribute their funds. *Health Research Policy and Systems, 14*, 12. https://doi.org/10.1186/s12961-015-0074-z

References

Ballabeni, A., Hemenway, D., & Scita, G. (2016). Time to tackle the incumbency advantage in science: A survey of scientists shows strong support for funding policies that would distribute funds more evenly among laboratories and thereby benefit new and smaller research groups. *EMBO Reports, 17*(9), 1254–1256. https://doi.org/10.15252/embr.201642998

Day, S., Wu, W., Mason, R., & Rochon, P. A. (2019). Measuring the data gap: Inclusion of sex and gender reporting in diabetes research. *Research Integrity and Peer Review, 4*, 9. https://doi.org/10.1186/s41073-019-0068-4

De Peuter, S., & Conix, S. (2022). The modified lottery: Formalizing the intrinsic randomness of research funding. *Accountability in Research, 29*(5), 324–345. https://doi.org/10.1080/08989621.2021.1927727

Diong, J., Kroeger, C. M., Reynolds, K. J., Barnett, A., & Bero, L. A. (2021). Strengthening the incentives for responsible research practices in Australian health and medical research funding. *Research Integrity and Peer Review, 6*(1), 11. https://doi.org/10.1186/s41073-021-00113-7

Gallo, S. A., Schmaling, K. B., Thompson, L. A., & Glisson, S. R. (2021). Grant review feedback: Appropriateness and usefulness. *Science and Engineering Ethics, 27*(2), 18. https://doi.org/10.1007/s11948-021-00295-9

Hankivsky, O., Springer, K. W., & Hunting, G. (2018). Beyond sex and gender difference in funding and reporting of health research. *Research Integrity and Peer Review, 3*, 6. https://doi.org/10.1186/s41073-018-0050-6

Haverfield, J., & Tannenbaum, C. (2021). A 10-year longitudinal evaluation of science policy interventions to promote sex and gender in health research. *Health Research Policy and Systems, 19*(1), 94. https://doi.org/10.1186/s12961-021-00741-x

Herbert, D. L., Barnett, A. G., Clarke, P., & Graves, N. (2013). On the time spent preparing grant proposals: An observational study of Australian researchers. *BMJ Open, 3*(5), e002800. https://doi.org/10.1136/bmjopen-2013-002800

Herbert, D. L., Coveney, J., Clarke, P., Graves, N., & Barnett, A. G. (2014). The impact of funding deadlines on personal workloads, stress and family relationships: A qualitative study of Australian researchers. *BMJ Open, 4*(3), e004462. https://doi.org/10.1136/bmjopen-2013-004462

Kalichman, M. (2020). Survey study of research integrity officers' perceptions of research practices associated with instances of research misconduct. *Research Integrity and Peer Review, 5*(1), 1–8. https://doi.org/10.1186/s41073-020-00103-1

Liu, M., Choy, V., Clarke, P., Barnett, A., Blakely, T., & Pomeroy, L. (2020). The acceptability of using a lottery to allocate research funding: A survey of applicants. *Research Integrity and Peer Review, 5*, 3. https://doi.org/10.1186/s41073-019-0089-z

Maisonneuve, H. (2019). Development of research integrity in France is on the rise: The introduction of research integrity officers was a progress. *Research Integrity and Peer Review, 4*, 20. https://doi.org/10.1186/s41073-019-0080-8

Mason, R. (2020). Doing better: Eleven ways to improve the integration of sex and gender in health research proposals. *Research Integrity and Peer Review, 5*(1), 15. https://doi.org/10.1186/s41073-020-00102-2

McCrabb, S., Mooney, K., Wolfenden, L., Gonzalez, S., Ditton, E., Yoong, S., & Kypri, K. (2021). "He who pays the piper calls the tune": Researcher experiences of funder suppression of health behaviour intervention trial findings. *PLoS One, 16*(8), e0255704. https://doi.org/10.1371/journal.pone.0255704

Morgan, B., Yu, L. M., Solomon, T., & Ziebland, S. (2020). Assessing health research grant applications: A retrospective comparative review of a one-stage versus a two-stage application assessment process. *PLoS One, 15*(3), e0230118. https://doi.org/10.1371/journal.pone.0230118

Ruppertsberg, A. I., Ward, V., Ridout, A., & Foy, R. (2014). The development and application of audit criteria for assessing knowledge exchange plans in health research grant applications. *Implementation Science, 9*(1), 1–6. https://doi.org/10.1186/s13012-014-0093-0

Scarrow, G., Angus, D., & Holmes, B. J. (2017). Reviewer training to assess knowledge translation in funding applications is long overdue. *Research Integrity and Peer Review, 2*, 13. https://doi.org/10.1186/s41073-017-0037-8

Senecal, J. B., Metcalfe, K., Wilson, K., Woldie, I., & Porter, L. A. (2021). Barriers to translational research in Windsor Ontario: A survey of clinical care providers and health researchers. *Journal of Translational Medicine, 19*(1), 479. https://doi.org/10.1186/s12967-021-03097-6

UK Research Integrity Office. (2022). *Code of Practice for Research*. https://ukrio.org/publications/code-of-practice-for-research/

Weber-Main, A. M., McGee, R., Eide Boman, K., Hemming, J., Hall, M., Unold, T., Harwood, E. M., Risner, L. E., Smith, A., Lawson, K., Engler, J., Steer, C. J., Buchwald, D., Jones, H. P., Manson, S. M., Ofili, E., Schwartz, N. B., Vishwanatha, J. K., & Okuyemi, K. S. (2020). Grant application outcomes for biomedical researchers who participated in the National Research Mentoring Network's Grant Writing Coaching Programs. *PLoS One, 15*(11), e0241851. https://doi.org/10.1371/journal.pone.0241851

Appendix 4.1

Survey about grant review performance

Section 1: Demographics

Question 1: What is your gender? Please choose ONLY ONE of the following:

- Male   ☐
- Female   ☐
- Prefer not to answer   ☐

Question 2: What is your age? Please choose ONLY ONE of the following:

- Under 30   ☐
- 30–39   ☐
- 40–49   ☐
- 50–59   ☐
- 60+   ☐

Question 3: Please specify your race/ethnicity Please choose ANY that apply:

- American Indian or Alaska Native   ☐
- Asian or Asian American   ☐
- Black or African American   ☐
- Hawaiian or Other Pacific Islander   ☐
- Hispanic or Latino Non-Hispanic   ☐
- White/Caucasian   ☐
- Other: _____
- Prefer not to answer   ☐

Question 4: What type of degree(s) do you have? Please choose ALL that apply:

- PhD or other research doctorate   ☐
- MD   ☐
- DDS   ☐
- DVM or VMD   ☐
- Other: _____
- Prefer not to answer   ☐

Question 5: What type of organisation do you work for? Please choose ONLY ONE of the following:

- Academia   ☐
- Government   ☐

- Industry □
- Other: _____

Question 6: What stage of career have you reached? Please choose ONLY ONE of the following:

- Early career □
- Mid-career □
- Late career/tenured □
- Emeritus □

Question 7: On average, how many hours do you work each week? Please choose ONLY ONE of the following:

- 40 hours □
- 40–50 hours □
- 50–60 hours □
- 60–70 hours □
- 70 + hours □

Please provide any comments that justify your responses under Section 1, Demographics. Please write your answer here:

_____
_____
_____

Section 2: Grant submission and peer review experience

Question 8: Have you submitted a grant for peer review in the last 3 years? Please choose ONLY ONE of the following:

- Yes □
- No □

    If you answered 'yes' to submitting a grant for peer review in the past 3 years, how many grant applications have you submitted in that time frame? Please choose ONLY ONE of the following:

- 1 □
- 2 □
- 3 □
- 4 □
- 5 □
- 6 □
- 7 or more □

Question 9: Did you receive reviewer feedback on your last grant submission? Please choose ONLY ONE of the following:

- Yes □
- No □

Question 10: Was your last application successful, that is, were you funded? Please choose ONLY ONE of the following:

- Yes ☐
- No ☐

Question 11: Have you served on a peer review panel in the last 3 years? Please choose ONLY ONE of the following:

- Yes ☐
- No ☐

If you answered 'yes' to serving on a peer review panel in the last 3 years, how many peer review panels have you served on in that time frame? Please choose ONLY ONE of the following:

- 1 ☐
- 2 ☐
- 3 ☐
- 4 ☐
- 5 ☐
- 6 ☐
- 7 or more ☐

Question 12: If you answered 'yes' to serving on a peer review panel in the past 3 years (i.e., Question 11), please select the mode of your last peer review panel meeting. Please choose ONLY ONE of the following:

- Face-to-face ☐
- Remote (video/teleconference) ☐
- Internet-assisted ☐
- Other ☐

Question 13: How many ad-hoc reviews (usually one or two grant applications reviewed telephonically that are being evaluated in a panel meeting setting) have you performed in the past 3 years? Please choose ONLY ONE of the following:

- 0 ☐
- 1 ☐
- 2 ☐
- 3 ☐
- 4 ☐
- 5 ☐
- 6 ☐
- 7 or more ☐

Question 14: Have you reviewed for a journal in the last 3 years? Please choose ONLY ONE of the following:

- Yes ☐
- No ☐

If you answered 'yes' to reviewing for a journal in the past 3 years, how many submissions have you reviewed in that time frame? Please choose ONLY ONE of the following:

- 1   ☐
- 2   ☐
- 3   ☐
- 4   ☐
- 5   ☐
- 6   ☐
- 7 or more   ☐

Question 15: What is a higher personal priority: grant review or journal review? Please choose ONLY ONE of the following:

- Grant review   ☐
- Journal review   ☐
- Both are equal priority   ☐
- Neither is a priority   ☐

Please elaborate on your responses under *Section 2, Grant submission and peer review experience*. Please write your answer here:

_____
_____
_____

Section 3: Investigator attitudes toward grant review

If you answered 'yes' to Question 9 (i.e., receiving feedback on your last grant submission) please answer Section 3 of the questionnaire. However, if you answered 'no' please proceed to Section 4.

Question 16: On a scale of 1–5 (1 most useful, 5 least useful), overall how useful was the reviewer feedback you received on your last grant submission? Please choose ONLY ONE of the following:

- 1 (Most useful)   ☐
- 2   ☐
- 3   ☐
- 4   ☐
- 5 (Least useful)   ☐

Question 17: On a scale of 1–5 (1 most useful, 5 least useful), how useful was the reviewer feedback in improving your grantsmanship? Please choose ONLY ONE of the following:

- 1 (Most useful)   ☐
- 2   ☐
- 3   ☐
- 4   ☐
- 5 (Least useful)   ☐

Question 18: If you were not funded, on a scale of 1–5 (1 most useful, 5 least useful), how useful was the reviewer feedback in improving your future submissions? Please choose ONLY ONE of the following:

- 1 (Most useful)   □
- 2   □
- 3   □
- 4   □
- 5 (Least useful)   □

Question 19: On a scale of 1–5 (1 most useful, 5 least useful), how useful was the reviewer feedback in informing your future scientific endeavours in the proposed research area? Please choose ONLY ONE of the following:

- 1 (Most useful)   □
- 2   □
- 3   □
- 4   □
- 5 (Least useful)   □

Question 20: Did you feel the reviewer feedback was well written, cohesive, and balanced? Please choose ONLY ONE of the following:

- Yes   □
- No   □

Question 21: Did you feel the reviewer feedback was fair and unbiased? Please choose ONLY ONE of the following:

- Yes   □
- No   □

Question 22: Overall, in what area(s) did the reviewer feedback primarily focus? Please choose ALL that apply:

- Potential impact of research   □
- Hypothesis   □
- Research methodology   □
- Innovation potential   □
- Preliminary data   □
- Responsiveness to funding mechanism   □
- Statistical issues   □
- Qualifications of research team   □
- Budget   □
- Other   □

Question 23: Did the reviewers comment on the riskiness of the research project? Please choose ONLY ONE of the following:

- Yes   □
- No   □

Question 24: Based on the reviewer feedback you received, do you feel that the reviewers had the appropriate expertise to evaluate your grant application? Please choose ONLY ONE of the following:

- Yes   □
- No   □

Please elaborate on your responses under Section 3, *Investigator attitudes towards grant review.* Please write your answer here:

_____
_____
_____

Section 4: Reviewer attitudes towards grant review

Question 25: What are your reasons for accepting an invitation to serve on a peer review panel? Please choose ALL that apply:

- Desire to give back to the scientific community   □
- Networking opportunities   □
- Informing your own grantsmanship   □
- Gaining exposure to new and innovative scientific areas   □
- Enhancing your career/resume   □
- Expectation from the funding agency   □
- Honorarium   □
- Other   □

Question 26: Do you feel that serving as a reviewer on peer review panels has positively impacted your career? Please choose ONLY ONE of the following:

- Yes   □
- No   □

Question 27: If you feel that serving as a peer reviewer has positively impacted your career, in what ways has serving as a reviewer influenced your career? Please choose ALL that apply:

- Bolstered your career   □
- Improved your grantsmanship   □
- Increased your exposure to new scientific ideas   □
- Improved your networking/collaboration opportunities   □
- Other: _____

Question 28: In general, which type of panel meeting format do you prefer? Please choose ONLY ONE of the following:

- Face-to-face   ☐
- Virtual [teleconference/videoconference]   ☐
- Internet-assisted   ☐

Question 29: On a scale of 1–5, (1 most influential, 5 least influential), please rate the following factors in influencing your selection of preferred panel meeting format: Please write your answer(s) here:

- Logistical convenience: _____
- Level of communication among panel members: _____
- Networking opportunities: _____
- Likelihood to participate on panel: _____

Question 30: In the last 3 years, how many times have you declined an invitation to serve on a peer review panel? Please choose ONLY ONE of the following:

- 1   ☐
- 2   ☐
- 3   ☐
- 4   ☐
- 5   ☐
- 6   ☐
- 7 or more   ☐

Question 31: What were your reasons for declining an invitation to serve on a peer review panel? Please choose ALL that apply:

- Limited free time   ☐
- Poor expertise match   ☐
- Personal reasons (holiday, sickness, travel)   ☐
- Review timeline too compressed   ☐
- Conflict of interest   ☐
- Issue with funding agency   ☐
- Other   ☐

Question 32: What is the maximum number of peer review panels/committees you prefer to serve on per year? Please choose ONLY ONE of the following:

- 1   ☐
- 2   ☐
- 3   ☐
- More than 3   ☐

Question 33: What is the maximum number of days you prefer to attend a peer review panel meeting? Please choose ONLY ONE of the following:

- 1   ☐
- 2   ☐
- 3   ☐
- More than 3   ☐

Question 34: What is the maximum number of grant applications you prefer to be assigned for a peer review panel meeting? Please choose ONLY ONE of the following:

- 1–2   ☐
- 3–4   ☐
- 5–6   ☐
- 7   ☐
- More than 7

Question 35: What was the actual number of days of your last peer review panel meeting? Please choose ONLY ONE of the following:

- 1   ☐
- 2   ☐
- 3   ☐
- More than 3   ☐

Question 36: What was the actual number of grant applications you were assigned to review at your last peer review panel meeting? Please choose ONLY ONE of the following:

- 1–2   ☐
- 3–4   ☐
- 5–6   ☐
- 7–8   ☐
- More than 8   ☐

Question 37: On average, how many hours did you spend reviewing each grant application before the panel meeting? Please choose ONLY ONE of the following:

- 1–2   ☐
- 2–3   ☐
- 3–4   ☐
- 4–5   ☐
- 5–6   ☐
- 7 or more   ☐

Please elaborate on your responses under *Section 4, Reviewer attitudes towards grant review.* Please write your answer here:

_____
_____
_____

Section 5: Peer review panel meeting proceedings

Question 38: Please answer the following questions in relation to your last peer review meeting. On a scale of 1–5 (1 most definitely, 5 not at all), was your scientific expertise necessary and appropriately used in the review process? Please choose ONLY ONE of the following:

- 1 (Most definitely)   □
- 2   □
- 3   □
- 4   □
- 5 (Not at all)   □

Question 39: On a scale of 1–5 (1 most definitely, 5 not at all), from your perspective was the expertise of the other panel members necessary and appropriately used in the review process? Please choose ONLY ONE of the following:

- 1 (Most definitely)   □
- 2   □
- 3   □
- 4   □
- 5 (Not at all)   □

Question 40: Did the grant application discussions facilitate reviewer participation? Please choose ONLY ONE of the following:

- Yes   □
- No   □

Question 41: Were the grant application discussions fair and balanced? Please choose ONLY ONE of the following:

- Yes   □
- No   □

Question 42: On a scale of 1–5 (1 most useful, 5 least useful), how useful were the grant application discussions in clarifying differing reviewer opinions? Please choose ONLY ONE of the following:

- 1 (Most useful)   □
- 2   □
- 3   □
- 4   □
- 5 (Least useful)   □

Question 43: On a scale of 1–5 (1 extremely effective, 5 no effect), did the grant application discussions affect the outcome? Please choose ONLY ONE of the following:

- 1 (Extremely effective)   □
- 2   □
- 3   □
- 4   □
- 5 (No effect)   □

Question 44: On a scale of 1–5 (1 most appropriate, 5 least appropriate), were the evaluation criteria appropriate to judge the best science and move the field forward? Please choose ONLY ONE of the following:

- 1 (Most appropriate) □
- 2 □
- 3 □
- 4 □
- 5 (Least appropriate) □

Question 45: On a scale of 1–5 (1 extremely important, 5 of no importance), how important is the PI's track record to assessing an investigator-initiated-type application? Please choose ONLY ONE of the following:

- 1 (Extremely important) □
- 2 □
- 3 □
- 4 □
- 5 (Of no importance) □

Question 46: In general, does a PI's track record temper your assessment of any detected methodological weaknesses? Please choose ONLY ONE of the following:

- Yes □
- No □

Question 47: On a scale of 1–5 (1 most definitely, 5 not at all), did the grant application discussions promote the best science? Please choose ONLY ONE of the following:

- 1 (Most definitely) □
- 2 □
- 3 □
- 4 □
- 5 (Not at all) □

Question 48: Was innovation factored into selecting the best science? Please choose ONLY ONE of the following:

- Yes □
- No □

Question 49: Did you view innovation as an essential component of scientific excellence when evaluating the grant applications? Please choose ONLY ONE of the following:

- Yes □
- No □

Question 50: Did the risks associated with innovative research impact the scores you assigned to the grant applications? Please choose ONLY ONE of the following:

- Yes ☐
- No ☐

Question 51: On a scale of 1–5 (1 completely, 5 not at all), how much did the seniority of your fellow panel members influence your evaluations during the panel deliberations? Please choose ONLY ONE of the following:

- 1 (Complete) ☐
- 2 ☐
- 3 ☐
- 4 ☐
- 5 (Not at all) ☐

Question 52: Was the format and duration of the grant application discussions sufficient to allow the non-assigned reviewers to cast well informed merit scores? Please choose ONLY ONE of the following:

- Yes ☐
- No ☐

Question 53: On a scale of 1–5 (1 extremely useful, 5 not useful at all), how useful was the Chair in facilitating the application discussions? Please choose ONLY ONE of the following:

- 1 (Extremely useful) ☐
- 2 ☐
- 3 ☐
- 4 ☐
- 5 (Not useful at all) ☐

Please elaborate on your responses under *Section 5, Peer review panel meeting proceedings.* Please write your answer here:

_____
_____
_____

Thank you for taking the time to fill out the survey. Have a wonderful day!

Source: Gallo, S. A., Schmaling, K. B., Thompson, L. A., & Glisson, S. R. (2021). Grant review feedback: Appropriateness and usefulness. *Science and Engineering Ethics, 27*(2), 18. https://doi.org/10.1007/s11948-021-00295-9

5 Improving equity, diversity, and inclusion in academia

Abstract

The purpose of this chapter is to present some main factors in academic workplaces that prevent the creation of high-quality research. It begins with a definition, prevalence, contributing factors, and solutions to bullying in academic workplaces. It then explains strategies to improve ethnic and gender diversity in research workplaces. By explaining these factors and associated solutions, it is anticipated that there will be improvements in the quality of research that is produced.

Keywords: Academic Bullying; Ethnic Diversity; Gender Diversity; Self-correcting Research Culture; Women in Academia

Key points

- Bullying, racism, and sexism can undermine the productivity and performance of researchers.
- Among other factors, bullying in research workplaces is caused by the pressure to create favourable citations metrics and the power differentials between junior and senior academics.
- Due to few networking opportunities and gender-based disparities, women are typically not on parity with males on editorial boards, senior academic roles, or first authors on manuscripts for some disciplines.
- Improving the participation of female academics and those from other minority backgrounds can ensure academic appointments are meritorious and that the academic workforce is more representative of the community that it serves.

5.1 Academic bullying

5.1.1 Overview of academic bullying

Academic bullying has been defined in different ways, can present itself in different forms, and can have different impacts on victims. Averbuch and colleagues (2021, p. 14) have defined it as 'The abuse of authority by the perpetrator who targets the victim in order to impede education or career growth through punishing behaviours that include overwork, destabilisation, and isolation in an academic setting'. Mahmoudi and Keashly (2021, p. 3338) have also defined academic bullying as 'Bullying in the academic workplace spans a wide spectrum of actions, some of which also occur in other work contexts such as: verbal abuse, public shaming, and isolation'. To quantify the extent, consequences, and reporting mechanisms for academic bullying

DOI: 10.4324/9781003510376-5

in medical settings, Averbuch and colleagues examined 68 studies that represented 82,349 respondents. They reported that the most common type of bullying behaviour identified was overwork (i.e., 35,779 respondents in 28 studies). The most common impact of academic bullying was psychological distress (i.e., 39.1% of all respondents in the studies reviewed). Furthermore, men were reported to often be the perpetrators of academic bullying (i.e., 67.2% of 4,722 respondents in five studies) and women were most commonly the victims of such bullying (i.e., 56.2% of 15,246 respondents in 27 studies). Only a few victims reported that they had been bullied (i.e., 28.9% of 9,410 victims in 25 studies) and most who did report such bullying did not believe that they had obtained a positive outcome (57.5%). A lack of enforcement of institutional policies (i.e., reported in 13 studies), hierarchical power structures (i.e., 7 studies), and normalisation of bullying (i.e., 10 studies) were the most common factors that contributed to the development of academic bullying.

5.1.2   *Factors that can cause and exacerbate academic bullying*

5.1.2.1   *Citation metrics*

During 2005 Jorge E. Hirsch, an Argentine-born American physicist, created the *h*-index. The *h*-index is intended to calculate a researcher's academic productivity and it has become a common feature on popular citation databases, such as Google Scholar, Scopus, and Web of Science. The *h*-index's continual usage can be attributed to two factors. First, a researcher's output can be summarised in a single number that enables an easy way to compare and rank scholars. Second, it is applicable to all scholars regardless of their career stage since a minimal number of publications and career experience is not required (Koltun & Hafner, 2021). Despite its benefits, some have criticised the *h*-index claiming that it is not an accurate way for judging a researcher's productivity (Hirsch & Buela-Casal, 2014). For example, in response to the question 'Can the h-index lead to unfair results?' Hirsch, the creator of the *h*-index, explained:

> I think it can, and should therefore be used with care. … some scientists publish relatively few papers but most of them have exceptional quality. This results in a relatively low h-index and an exceptionally high number of citations. Other scientists may have a high h-index because they collaborate with influential scientists, while not being themselves the creative driving force in the research. Scientists who conduct research on subjects that are more "fashionable" will have higher h-indices even if they are not necessarily better than other scientists who work on profound questions and write papers that may have a lasting but not immediate impact. Scientists that publish in large collaborations will have larger h-indices than those publishing alone or with few coauthors. Each case is different, which is why in addition to the h-index and other bibliometric indicators it is important to consider the totality of the scientist, read his/her papers and consider his/her production beyond the published papers as well as his/her reputation among their peers, to obtain a comprehensive evaluation.
>
> (Hirsch & Buela-Casal, 2014, pp. 163–164)

Hirsch's concerns have also been reiterated by other scholars. Mahmoudi and colleagues (2020) explained two negative consequences of the *h*-index. First, to achieve a large *h*-index score researchers may abandon their research interests so that they can focus on writing articles about topics that are more popular and more likely to be cited. Second, to increase their *h*-index scores, researchers often form collaborations to produce articles that make no significant or meaningful contributions, but which increase each author's *h*-index (Mahmoudi et al., 2020). Adding to these concerns, Ashrafi-rizi and colleagues (2021) have claimed that *h*-index scores can create anxiety among researchers since they are used to determine career progression.

To mitigate the negative consequences caused by *h*-index scores, such as the anxiety and stress that some scholars feel about their *h*-index scores and the competitive bullying between scholars, the *h*-index itself should not be viewed in isolation when deciding an academic's career progression. Instead, a scholar's *h*-index should be one aspect when deciding both their career progression and awarding of grants. Other aspects that should be considered is student reports about teaching performance, research visibility as determined by altmetric scorers, and ability to collaborate with other leading scholars and research institutions. Such a broad perspective can illustrate the researcher's ability to collaborate with others and produce research that is greatly disseminated through being visible.

5.1.3   *Strategies to reduce academic bullying*

5.1.3.1   *Legislative changes*

Currently, victims of discrimination and sexual harassment have institutional reporting systems and legal remedies which support them in finding justice for these crimes. However, victims of academic bullying, who suffer similar repercussions, have no legal or institutional remedies. Because academic bullying is not a crime, targets often suffer in silence because there is no recourse.

(Mahmoudi & Moss, 2020, p. 1)

As Mahmoudi and Moss (2020) explained above, within some nations, there are insufficient legal safeguards that deter academics from bullying other academic staff or protect the victims of academic bullying. This legal void has created debate about if research misconduct should be criminalised (Bülow & Helgesson, 2019; Dal-Ré et al., 2020). Regardless of this commentary, due to a lack of legal protections, those who experience academic bullying are inclined to resign. When such resignations occur the talent and diversity of the academic workforce diminishes along with the quality of research that is produced. To prevent this situation from occurring, legislation that protects the victims of academic bullying and punishes the perpetrators of such acts needs to be created and enforced.

5.2   Racism in research workplaces

For the best possible research to be created all suitably qualified and experienced academic candidates should be appointed to academic positions regardless of their immutable characteristics, such as their age, gender identity and expression, and ethnicity. Despite this meritocratic view, only a few culturally and ethnically diverse academics occupy influential positions. The lack of ethnic diversity within science has been acknowledged (Cell Editorial Team, 2020). Furthermore, this lack of diversity is acknowledged in other academic fields, such as the editorial board of the journal *Autism: The International Journal of Research and Practice* (Jones & Mandell, 2020).

To improve the ethnic diversity within researcher teams several opinion pieces have outlined strategies to can be implemented to curb racism in research workplaces and promote the inclusion of ethnic minorities (Arif et al., 2021; Chaudhary & Berhe, 2020; Flores et al., 2019; Massey et al., 2022). Arif and colleagues (2021) have published ten rules for supporting students who have historically been underrepresented in science. Similarly, Chaudhary and Berhe (2020) wrote an opinion piece about ten rules that can be used to create anti-racist laboratories, these rules were:

- Rule 1: Lead informed discussions about anti-racism in your lab regularly
- Rule 2: Address racism in your lab and field safety guidelines
- Rule 3: Publish papers and write grants with Black, Indigenous, and people of colour (BIPOC) colleagues
- Rule 4: Evaluate your lab's mentoring practices
- Rule 5: Amplify voices of BIPOC scientists in your field
- Rule 6: Support BIPOC in their efforts to organise
- Rule 7: Intentionally recruit BIPOC students and staff
- Rule 8: Adopt a dynamic research agenda
- Rule 9: Advocate for racially diverse leadership in science
- Rule 10: Hold the powerful accountable and don't expect gratitude

To assist with practically implementing these strategies, Miller-Kleinhenz and colleagues (2021) have published a training curriculum that can be used to educate research staff about how to handle racism in academia (see Appendix 5.1). Underpinning their curriculum is the concept that conversations between academics is required to build an academic workplace that is anti-racist and equitable (see Figure 5.1). Additionally, there are five guiding principles for having productive and continuous conversations about race and social justice in academia (see Table 5.1).

A.

B.

Step One:
participate in ongoing conversations
with your peers, listen and learn

Step Two:
work together to build an academic
environment that is welcoming

Figure 5.1 A model for building an equitable environment for academics through conversation

Source: Miller-Kleinhenz et al. (2021, p. 3).

Note: On the left-hand side, a diverse group of people engages in a conversation while having a picnic. On the right-hand side, a diverse group of people works together to build a home representing academia, with different colours of wooden planks, representing the many skin tones that are present in academia.

Table 5.1 Guiding principles for having productive, long-term conversations around issues of race and social justice in academia

Conscious effort leads to sustainable action and change.	*Dismantling racism in academia is not a quick fix. Anti-racism requires a time commitment and willingness to work towards a lasting change.*
Building trust allows for open and honest conversations.	Everyone in the group must feel psychologically safe in order to freely share their feelings without fear of retribution. Conversations around issues of race and social justice can start within small groups and cohorts, which helps provide the foundation and trust needed for the difficult and honest conversations throughout the course. In academia, there are already such 'cohorts' or bonded groups in the form of departments, labs, committees, class groups, etc.
Being comfortable having uncomfortable conversations facilitates growth.	Conversations about race are not easy. Often, we dampen or avoid the ugly truth to maintain a semblance of harmony or to hide personal biases. However, tackling institutional racism is impossible without first recognising and tackling our own biases, acknowledging that people are the source of widespread institutional racism, and all people have the capacity to change and do better.
Active listening allows others to feel that they are seen, and their experiences are heard.	Listening is not just waiting to speak; it is hearing what others have to say and giving them the space to share their experiences without being dismissive or inserting our personal ego. This includes framing questions and discussions in ways that respect others to prevent gaslighting.
Empathy towards others' experiences validates their lived experience.	You may have never experienced racism, but you can empathise with those who have. Avoid dismissing the experiences of the oppressed or arguing on behalf of the oppressor.
Diversity is a strength.	Diversity within the group will lead to the diversity of thoughts. Given that the topics for discussion/reflection are decided by the group, the more a group can leverage its diversity, the more meaningful the conversations will be.
Engaging and becoming agents of change advances anti-racism everywhere.	While conversations around race may initially take place within a cohort, the ultimate goal is to extend them into our various communities. Conversations on race, equity, and social justice can be uncomfortable and the natural reaction is to withdraw and disengage. However, constant engagement, and encouraging the same engagement from peers, students, friends, and family will allow all of us to move past the discomfort and into growth. This domino effect will help spur others into action.

Source: Miller-Kleinhenz et al. (2021, p. 5).

5.3   Women in academia

5.3.1   *Women's participation in academia*

5.3.1.1   *Representation of women in senior academic roles*

In comparison to their male counterparts, fewer women are employed in senior academic ranks (Jamorabo et al., 2021; Linscheid et al., 2020; Liu et al., 2021). For example, of the 3,655 faculty positions examined within 163 academic gastroenterology programmes in the United States of America, Jamorabo and colleagues (2021) reported that women comprised 1,049 (28.7%) of the cohort. Additionally, a minority of women occupied faculty positions (26.8%) and fellowship programme directors and divisional/departmental chairs and chiefs (19.4%). In another example, Linscheid and colleagues (2020) used personnel data from the *American Association of Medical College* to measure the proportion of female academic faculty representation for surgical specialities from 1969 to 2018. As illustrated below, compared to general surgery the overall representation of women in academic surgery is slowly increasing in orthopaedic surgery, neurosurgery, cardiothoracic surgery, and urology (see Figure 5.2).

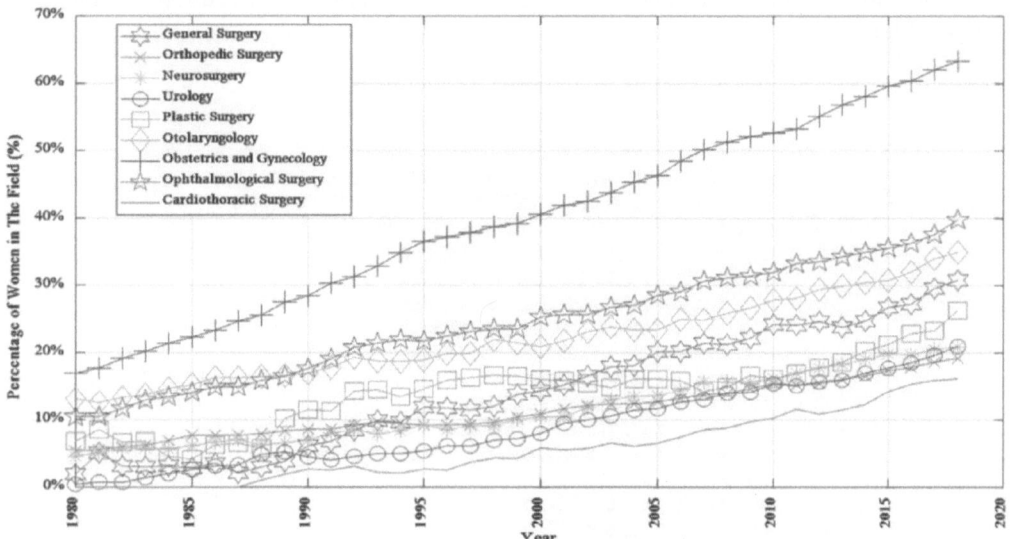

Figure 5.2 Percentage of women among practicing academic surgeons from 1980 to 2018

Source: Linscheid et al. (2020, p. 3).

5.3.1.2   Representation of women on editorial boards

Regarding the gender composition of editorial boards women are not on parity with men (Grinnell et al., 2020; Lobl et al., 2020; O'Brien et al., 2021; Pinho-Gomes et al., 2021). Grinnell and colleagues (2020) measured the proportion of women in editorial boards in women's health journals. Ironically, despite being women's health issues, of the 1,440 board members examined, 602 (42%) were women and 838 (58%) were men. In addition, women held 54 of 132 editor-in-chief positions (41%), 257 of 596 associated positions (43%), 13 of 42 deputy editor positions (30%), 46 of 120 section editor positions (38%), and 232 of 549 other editor positions (42%). Based on these and other findings, Grinnell and colleagues concluded that women are underrepresented at most editorial levels in women's health journals, particularly journals about reproductive health, obstetrics and gynaecology, perinatology, gynaecological oncology, and breastfeeding.

The lack of women's representation on editorial boards has also been measured in the discipline of medicine and dermatology (Lobl et al., 2020). Lobl and colleagues (2020) conducted a search for dermatology journals on Medline Journal Citation Reports, Scopus, and Embase. The number and percentage of male and female editors in four different positions (i.e., Editor-in-Chief, Deputy Editor, Editorial Board, and other editorial staff) were measured for each journal's editorial board. They reported that 18% of editor-in-chief positions, 36% of deputy editor roles, 27% of overall editorial board positions, and 22% of other board roles were held by women. They concluded that women were underrepresented as editors at all levels in dermatology journals. Pinho-Gomes and colleagues (2021) investigated the gender ratio of male and female editors-in-chief in leading medical journals. They reported that five medical disciplines did not have any women as editors-in-chief (i.e., dentistry, oral surgery, and medicine; allergy; psychiatry; anaesthesiology; and ophthalmology). However, primary health care, microbiology, and genetics and heredity had more women as editors-in-chief than men (see Figure 5.3).

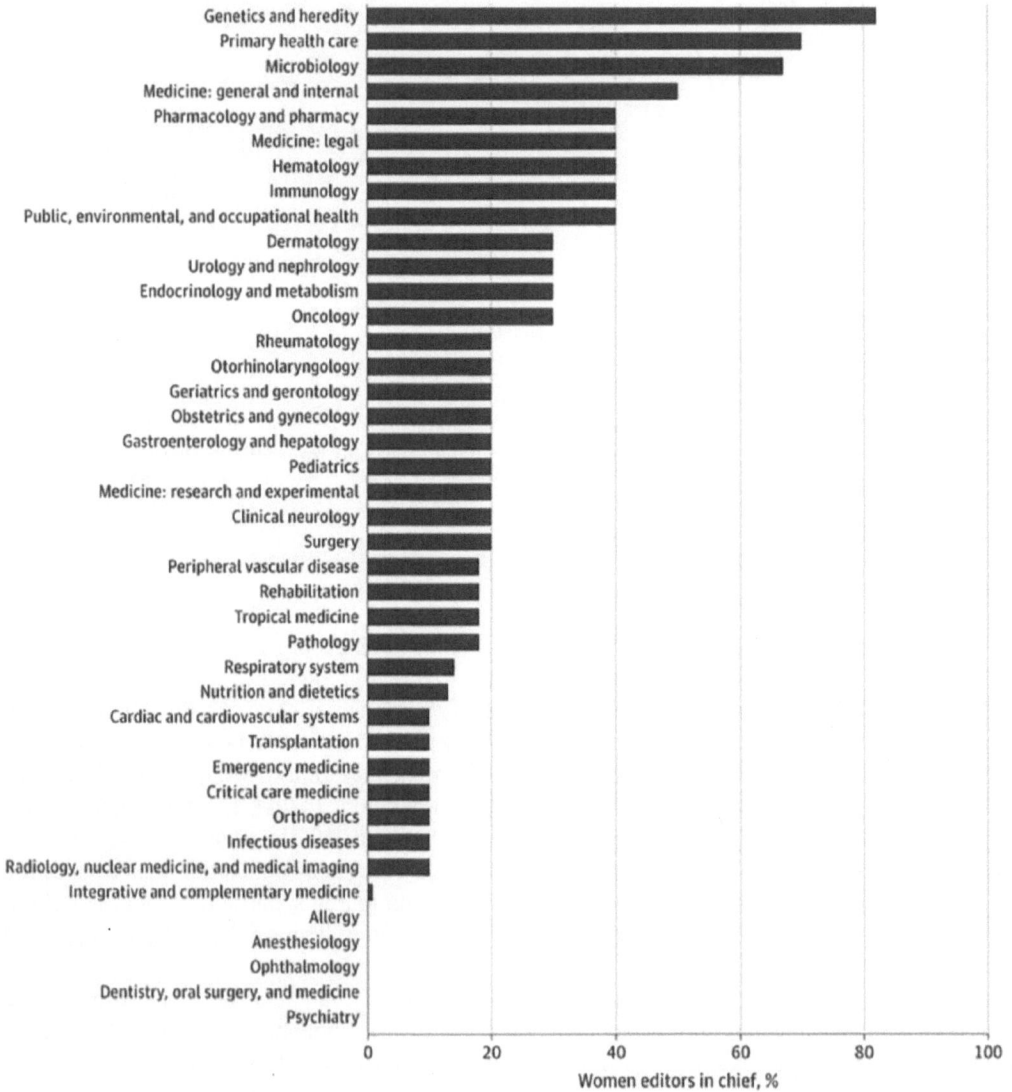

Figure 5.3 Percentage of women as editors-in-chief in the top ten highest-impact factor journals in each category of medical journals

Source: Pinho-Gomes et al. (2021, p. 3).

5.3.2   *Factors that inhibit the inclusion and promotion of women in academia*

5.3.2.1   *Wage disparities based on gender*

The wage disparities based on gender can dissuade women from applying for academic positions. Miller and colleagues (2022) reported that total monetary compensation was significantly greater for men across all professional ranks in both general surgery and obstetrics and gynaecology. Additionally, women faculty members within these departments earned almost $75,000 less than their male colleagues due to receiving more supplemental income and specialist allowances. Such wage disparities can discourage women from applying for, or remaining within, academic positions.

5.3.2.2   *Networking opportunities*

The lack of professional networking opportunities can inhibit women progressing into more senior roles in academia. Exploring this topic, Murphy and colleagues (2022) conducted in-depth, semi-structured interviews with 52 women and 52 men who worked in medical faculties at 16 institutions across the United States of America during 2019. An examination of the interview transcripts revealed that informal networks among male academics accelerated their career progression, often at the detriment of female academics. This sentiment was described the following way by Murphy and colleagues (2022, p. 8):

> Through gender-exclusive networking activities, such as golf, men established and maintained relationships with men faculty peers and built professional relationships with men in leadership positions or men from other departments or institutions. These relationships yielded different types of benefits, including solidarity within their unit, research assignments, letters of recommendation, robust networks within and across institutions, and direct access to people in leadership positions. These benefits translated into concrete professional advantages such as opportunities to publish, weigh in on important decisions, garner letters of recommendation from colleagues outside of one's home institution, and nominations to professional societies. These advantages then translated into measurable professional accomplishments, such as publications or promotions or membership in professional societies, which are ultimately presumed to be reflections of individual merit – rather than a function of one's professional network.

5.3.3   *Improving the inclusion, retention, and promotion of women in academia*

Strategies that promote the inclusion and career progression of women in academia have been explored (Cardel et al., 2020). Cardel and colleagues (2020) have outlined several strategies to help the inclusion, retention, and promotion of women in academia. First, to decrease recruitment bias, and maximise the prospect of the most suitably qualified person, regardless of gender, being appointed at least two female candidates should be interviewed for the advertised academic position. Second, academic and research institutions should improve their 'family friendly' policies to attract academic personnel who have family or caring responsibilities. Examples of such policies include access to childcare, implement clear application procedures for paid leave policies, and foster family friendly work events. Third, women academics should have access to faculty mentors who can have a positive influence on their careers. Fourth, academic and research institutions should implement and enforce policies that reduce any biases based on performance or the allocation of tasks. By implementing such strategies, the possible inclusion of the best possible candidate for an academic position, regardless of biological sex or gender expression, would be achieved.

5.4   General recommendations for improving equity, diversity, and inclusion in academia

5.4.1   Changes to publishing policies

5.4.1.1   Adopt a journal diversity statement with clear, actionable steps to achieve it

Excellence and creativity in scientific publications are enhanced by including more diverse perspectives. A practical first step is the adoption of a journal diversity statement with clear and doable actions. It is easier to guarantee that everyone understands the issue when it is explicitly defined. Additionally, this process involves top leadership demonstrating to writers, reviewers, and editors that equity, diversity, and inclusion principles are a priority. According to several reports, these award programmes serve as a catalyst for action about equity, diversity, and inclusion, which results in workplaces that are more welcoming. Over 47 publishing companies have established equity, diversity, and inclusion programmes. For example, Wiley Publishing has created guidelines to help editors create an equity, diversity, and inclusion statement. There are three steps in their procedure: (1) evaluating the needs of the journal and the research community, (2) identifying the journal's top priorities for action (e.g., altering the recruitment procedure and increasing the diversity of invited reviewers), and (3) creating an active statement that recognises that this process is ongoing and will call for periodic review to address unanswered questions (Dewidar et al., 2022).

5.4.1.2   Promote the use of inclusive and bias-free language

It may be easier to attract members of underrepresented groups if publishing practices don't perpetuate prejudice or disparaging views. In response, journals should practice using inclusive and prejudice-free language in all written communications and on the journal's website. Editors should speak to people or groups according to how they like to express themselves, their experiences, and their activities considering language changes through time. For example, the advice to use the singular "they" to refer to people when their identifiable pronouns are unclear is an addition to the 7th edition of the American Psychological Association's style guide. In another example, when the language of advertised fellowships was modified for gender inclusion, the University of Nottingham found an increase in the recruitment of female researchers in science, technology, engineering, and mathematics fields (Dewidar et al., 2022).

5.4.1.3   Appoint a journal's equity, diversity, and inclusion director or lead

When leaders utilise the authority that comes with their roles to campaign for equity, diversity, and inclusion, it may assist others in their efforts to eliminate prejudice and discrimination. Editors-in-chief should prioritise placing researchers from marginalised backgrounds who have equity, diversity, and inclusion competence into positions of leadership and advocacy. They could, albeit less preferably, designate one of their associate editors who comes from an underrepresented background or hire someone with equity, diversity, and inclusion knowledge who does not come from an underrepresented background. It would be prudent to form a consultative group for this leader that includes underrepresented academics, equity, diversity, and inclusion experts, and members of the public who have unique, lived experiences. The viewpoints of people from minorities may be critical for the team's success since their input will generate more culturally competent and practical solutions. The lead's tasks might include examining journal procedures while working with the editors-in-chief, promoting awareness of unconscious prejudice among editorial teams, and implementing measures to enhance equity, diversity, and inclusion. The lead should also oversee devising ways to diversity the editorial teams, peer reviewers, and authorship, as well as monitoring the journal's progress towards equity, diversity, and inclusion. The person or group in charge of this appointment should examine the journal's sources for hiring new staff members as well as the linguistics of its invitations to join editorial teams. It should be noted that knowledge of equity, diversity, and inclusion concepts and expertise in the sector are not enough to accomplish these objectives. Leaders who want to fill this position should be capable of conceiving plans that complement the goals and resources of the journal (Dewidar et al., 2022).

5.4.1.4   Establish a mentoring approach

There is plethora of evidence that in editorial positions certain populations are underrepresented. This limits their capacity to get the necessary expertise to assume leadership roles. Finding editorial board members in all fields is difficult, therefore finding editors with varied backgrounds, gender identities, races, and geographical regions would be much more difficult. A varied and representative team, on the other hand, may be more likely to demonstrate improved cultural competency due to their more diverse collection of life experiences. There should be efforts made to assemble a diverse staff, and any shortcomings in diversity should be openly acknowledged as a work in progress. Additionally, all editorial positions should have a set term since anyone in a position of authority for an extended period is likely to perpetuate disparities (Dewidar et al., 2022).

Instead of relying primarily on personal networks, journals might make open calls for reviewer positions to increase the variety of their reviewer pool. To maximise inclusion, it is important to evaluate the language and posting places of these adverts. The use of artificial intelligence (AI) or algorithms to select reviewers should be used with caution since they have the potential to reinforce bias against academics from low- and middle-income countries and marginalised groups. Thus, editors should be vigilant when using AI and correct any biases that may arise from its use. Additionally, journal editors may solicit suggestions from authors about potential reviewers from backgrounds that are underrepresented. It could be advantageous to invite reviewers who are familiar with the substance of the article when sending out invitations. Knowing the author's name, organisation, standing in the profession, or location may cause unconscious bias and compromise the neutrality of the peer review process. Journal editors should consider a double anonymous peer review procedure, in which the peer reviewers are unaware of the authors of the paper and vice versa, to reduce unconscious bias (Dewidar et al., 2022).

Establishing a mentorship programme may be a practical strategy to help inexperienced candidates for journal posts become prepared for the position in the future. Senior editorial team members might connect with more junior team members and customise their mentorship to meet the junior team member's requirements. Mentors should undertake unconscious bias training or other equity, diversity, and inclusion training as necessary (i.e., microaggressions, anti-racism) before participating in mentorship activities because they most likely come from non-underrepresented groups. Given that most editorial jobs are volunteer, mentorship initiatives must be supported by praise and encouragement. Mentors might be compensated for their time, or internal prizes for mentor performance could be established, which could aid in their promotions and tenure. The adoption of these tactics by numerous publications may aid in the formation of a network of mentors from whom they could be recruited for mentoring activities.

Training in research integrity may help them to support their mentees and fostering a supportive atmosphere. VIRT2UE Train is designed for anyone who wants to become research integrity trainers. The training focuses on creating and implementing high moral standards and behaviours based on the European Code of Conduct for Research Integrity (Dewidar et al., 2022).

5.4.1.5   *Monitor adherence to equity, diversity, and inclusion principles*

Meaningful and precise data about the makeup of editors, peer reviewers, and writers is necessary to discover gaps in diversity. To effectively assess journal success and adjust their aims, journal editors must routinely gather demographic data. The research community should be given a standard list of questions so they may freely offer self-identification information on things like career stage, gender, colour and ethnicity, and location. Journals can use the eight identifying categories suggested in the questionnaire provided by the Employment Equity Act as a starting point and make the necessary adjustments. As an alternative, journals may use outside services, like TOP factor, to track journal metrics when implementing equity, diversity, and inclusion policies (Dewidar et al., 2022).

5.4.1.6   *Publish reports on equity, diversity, and inclusion*

Journals and publishers should make their diversity statistics public so that they can be held accountable for their progress. As a result, when collecting self-identifiable information from participants, journals should ensure that they have informed consent. Their data should be handled with extreme caution and preserved with extreme care. To protect the confidentiality of participants, journals should only publish aggregated data (Dewidar et al., 2022).

5.4.2   **The role of universities and academic institutions**

When it comes to standard measurements and assessment methods in academia, students from underrepresented groups encounter many challenges. Heller and colleagues (2014) discovered that when grade point average requirements increased in medical schools in the United States of America from 2005 to 2009, the diversity of students in medical courses declined. This shows that assessments that rely mainly on academic indicators frequently neglect equity, diversity, and inclusion. As a result, developing a new concept of student academic achievement may aid in the improvement of equity, diversity, and inclusion at academic institutions (Dewidar et al., 2022).

Several ways of increasing diversity among trainees and early-career researchers have been successful. However, disparities in recruitment, retention, and promotion rates across numerous categories such as age, gender, and race must be addressed. This might be attributable, in part, to the limiting reliance on citation metrics and publications for evaluating these processes. Institutions should provide effective mentorship to underrepresented populations and integrate tenure or promotion assessments in recruiting. These honours include the Presidential Award for Excellence in Science, Math, and Engineering Mentoring from the National Science Foundation and the Eureka Award from the Australian Museum. Extending the definition of success to incorporate non-academic variables will improve the selection of diverse candidates and lay the groundwork for future success (Dewidar et al., 2022).

Academic course coordinators should also think about delivering the curriculum from an EDI viewpoint by diversifying the readings for their courses as well as the sources of the research that went into creating the course materials. The inclusion of diverse students, faculty, and relevant subjects are encouraged by placing a strong emphasis on diversity in the curriculum.

Additionally, underrepresented groups are more effectively engaged by curricula that mirror their actual experiences (Dewidar et al., 2022).

5.4.3   *The role of funding agencies*

The significance of equity research has been recognised by several funding organisations, including the National Institutes of Health and the Canadian Institutes of Health Research. Since diverse teams create more inventive, creative, and significant science, including diversity elements as a criterion for scoring may improve decisions about what research should be funded. Financial organisations might also develop grants that only disadvantaged academics can apply for, such opportunities give them greater chances to participate in the creation of research and perhaps eliminate any funding discrepancies in research. Examples of such funding opportunities include the Mental Health Dissertation Research Grant to Increase Diversity funded by the National Institute of Health and the Louis Stokes Alliances for Minority Participation funded by the National Science Foundation. Grant organisations might also consider mandating a certain proportion of researchers from underrepresented groups to serve as reviewers on funding panels. Such acts though might result in a 'diversity tax', putting a burden on underrepresented researchers. It is important to emphasise, however, that this 'diversity tax' only becomes problematic when serving on funding panels are not career-enhancing (Dewidar et al., 2022).

5.5   Conclusion

The main message of this chapter was that the discrimination of marginalised groups in academia, such as women and people of colour, has prevented them from achieving their potential as researchers. It began with a description about bullying in academia, including definitions, factors that cause bullying behaviours, and strategies to prevent bullying. It then discussed racism and discrimination against women in academia. To reduce the occurrence of these issues changes in university environments, changes in publication policies, and strategies that funding agencies can do were canvassed.

Additional readings

Bartlett, M. J., Arslan, F. N., Bankston, A., & Sarabipour, S. (2021). Ten simple rules to improve academic work-life balance. *PLoS Computational Biology, 17*(7), e1009124. https://doi.org/10.1371/journal.pcbi.1009124

Dewidar, O., Elmestekawy, N., & Welch, V. (2022). Improving equity, diversity, and inclusion in academia. *Research Integrity and Peer Review, 7*(1), 4. https://doi.org/10.1186/s41073-022-00123-z

Golden, N., Devarajan, K., Balantic, C., Drake, J., Hallworth, M. T., & Morelli, T. L. (2021). Ten simple rules for productive lab meetings. *PLoS Computational Biology, 17*(5), e1008953. https://doi.org/10.1371/journal.pcbi.1008953

Haven, T. L., Tijdink, J. K., Pasman, H. R., Widdershoven, G., Ter Riet, G., & Bouter, L. M. (2019). Researchers' perceptions of research misbehaviours: A mixed methods study among academic researchers in Amsterdam. *Research Integrity and Peer Review, 4*, 25. https://doi.org/10.1186/s41073-019-0081-7

Hofmann, B., Bredahl Jensen, L., Eriksen, M. B., Helgesson, G., Juth, N., & Holm, S. (2020). Research integrity among PhD students at the faculty of medicine: A comparison of three Scandinavian Universities. *Journal of Empirical Research on Human Research Ethics: JERHRE, 15*(4), 320–329. https://doi.org/10.1177/1556264620929230

Lewitter, F., Bourne, P. E., & Attwood, T. K. (2019). Ten simple rules for avoiding and resolving conflicts with your colleagues. *PLoS Computational Biology, 15*(1), e1006708. https://doi.org/10.1371/journal.pcbi.1006708

Naidoo-Chetty, M., & du Plessis, M. (2021). Job demands and job resources of academics in higher education. *Frontiers in Psychology, 12*, 631171. https://doi.org/10.3389/fpsyg.2021.631171

References

Abdalla, M., Abdalla, M., Abdalla, S., Saad, M., Jones, D. S., & Podolsky, S. H. (2023). The underrepresentation and stagnation of female, Black, and Hispanic authorship in the Journal of the American Medical Association and the New England Journal of Medicine. *Journal of Racial and Ethnic Health Disparities, 10*(2), 920–929. https://doi.org/10.1007/s40615-022-01280-z

Arif, S., Massey, M., Klinard, N., Charbonneau, J., Jabre, L., Martins, A. B., Gaitor, D., Kirton, R., Albury, C., & Nanglu, K. (2021). Ten simple rules for supporting historically underrepresented students in science. *PLoS Computational Biology, 17*(9), e1009313. https://doi.org/10.1371/journal.pcbi.1009313

Ashrafi-Rizi, H., Kazempour, Z., & Khazaie, S. (2021). H-index anxiety among health researchers: A commentary. *International Journal of Preventive Medicine, 12*, 146. https://doi.org/10.4103/ijpvm.IJPVM_598_20

Averbuch, T., Eliya, Y., & Van Spall, H. (2021). Systematic review of academic bullying in medical settings: Dynamics and consequences. *BMJ Open, 11*(7), e043256. https://doi.org/10.1136/bmjopen-2020-043256

Bülow, W., & Helgesson, G. (2019). Criminalization of scientific misconduct. *Medicine, Health Care, and Philosophy, 22*(2), 245–252. https://doi.org/10.1007/s11019-018-9865-7

Cardel, M. I., Dhurandhar, E., Yarar-Fisher, C., Foster, M., Hidalgo, B., McClure, L. A., Pagoto, S., Brown, N., Pekmezi, D., Sharafeldin, N., Willig, A. L., & Angelini, C. (2020). Turning chutes into ladders for women faculty: A review and roadmap for equity in academia. *Journal of Women's Health (2002), 29*(5), 721–733. https://doi.org/10.1089/jwh.2019.8027

Carter, A. J., Croft, A., Lukas, D., & Sandstrom, G. M. (2018). Women's visibility in academic seminars: Women ask fewer questions than men. *PLoS One, 13*(9), e0202743. https://doi.org/10.1371/journal.pone.0202743

Cell Editorial Team. (2020). Science has a racism problem. *Cell, 181*(7), 1443–1444. https://doi.org/10.1016/j.cell.2020.06.009

Chatterjee, P., & Werner, R. M. (2021). Gender disparity in citations in high-impact journal articles. *JAMA Network Open, 4*(7), e2114509. https://doi.org/10.1001/jamanetworkopen.2021.14509

Chaudhary, V. B., & Berhe, A. A. (2020). Ten simple rules for building an antiracist lab. *PLoS Computational Biology, 16*(10), e1008210. https://doi.org/10.1371/journal.pcbi.1008210

Cofnas, N. (2016). Science is not always "self-correcting". *Foundations of Science, 21*(3), 477–492. https://doi.org/10.1007/s10699-015-9421-3

Dal-Ré, R., Bouter, L. M., Cuijpers, P., Gluud, C., & Holm, S. (2020). Should research misconduct be criminalized? *Research Ethics, 16*(1–2), 1–12. https://doi.org/10.1177/1747016119898400

Dewidar, O., Elmestekawy, N., & Welch, V. (2022). Improving equity, diversity, and inclusion in academia. *Research Integrity and Peer Review, 7*(1), 4. https://doi.org/10.1186/s41073-022-00123-z

Flores, G., Mendoza, F. S., DeBaun, M. R., Fuentes-Afflick, E., Jones, V. F., Mendoza, J. A., Raphael, J. L., & Wang, C. J. (2019). Keys to academic success for under-represented minority young investigators: Recommendations from the Research in Academic Pediatrics Initiative on Diversity (RAPID) National Advisory Committee. *International Journal for Equity in Health, 18*(1), 93. https://doi.org/10.1186/s12939-019-0995-1

Forsberg, E. M., Anthun, F. O., Bailey, S., Birchley, G., Bout, H., Casonato, C., Fuster, G. G., Heinrichs, B., Horbach, S., Jacobsen, I. S., Janssen, J., Kaiser, M., Lerouge, I., van der Meulen, B., de Rijcke, S., Saretzki, T., Sutrop, M., Tazewell, M., Varantola, K., Vie, K. J., … Zöller, M. (2018). Working with research integrity-guidance for research performing organisations: The Bonn PRINTEGER statement. *Science and Engineering Ethics, 24*(4), 1023–1034. https://doi.org/10.1007/s11948-018-0034-4

Grinnell, M., Higgins, S., Yost, K., Ochuba, O., Lobl, M., Grimes, P., & Wysong, A. (2020). The proportion of male and female editors in women's health journals: A critical analysis and review of the sex gap. *International Journal of Women's Dermatology, 6*(1), 7–12. https://doi.org/10.1016/j.ijwd.2019.11.005

Heller, C. A., Rúa, S. H., Mazumdar, M., Moon, J. E., Bardes, C., & Gotto, A. M., Jr (2014). Diversity efforts, admissions, and national rankings: Can we align priorities? *Teaching and Learning in Medicine, 26*(3), 304–311. https://doi.org/10.1080/10401334.2014.910465

Hirsch, J. E., & Buela-Casal, G. (2014). The meaning of the h-index. *International Journal of Clinical and Health Psychology, 14*(2), 161–164. https://doi.org/10.1016/S1697-2600(14)70050-X

Jamorabo, D. S., Chen, R., Gurm, H., Jahangir, M., Briggs, W. M., Mohanty, S. R., & Renelus, B. D. (2021). Women remain underrepresented in leadership positions in academic gastroenterology throughout the United States. *Annals of Gastroenterology, 34*(3), 316–322. https://doi.org/10.20524/aog.2021.0597

Jones, D. R., & Mandell, D. S. (2020). To address racial disparities in autism research, we must think globally, act locally. *Autism: The International Journal of Research and Practice, 24*(7), 1587–1589. https://doi.org/10.1177/1362361320948313

Koltun, V., & Hafner, D. (2021). The h-index is no longer an effective correlate of scientific reputation. *PLoS One, 16*(6), e0253397. https://doi.org/10.1371/journal.pone.0253397

Linscheid, L. J., Holliday, E. B., Ahmed, A., Somerson, J. S., Hanson, S., Jagsi, R., & Deville, C., Jr. (2020). Women in academic surgery over the last four decades. *PLoS One, 15*(12), e0243308. https://doi.org/10.1371/journal.pone.0243308

Liu, X., Dunlop, R., Allavena, R., & Palmieri, C. (2021). Women representation and gender equality in different academic levels in veterinary science. *Veterinary Sciences, 8*(8), 159. https://doi.org/10.3390/vetsci8080159

Lobl, M., Grinnell, M., Higgins, S., Yost, K., Grimes, P., & Wysong, A. (2020). Representation of women as editors in dermatology journals: A comprehensive review. *International Journal of Women's Dermatology, 6*(1), 20–24. https://doi.org/10.1016/j.ijwd.2019.09.002

Mahmoudi, M., Ameli, S., & Moss, S. (2020). The urgent need for modification of scientific ranking indexes to facilitate scientific progress and diminish academic bullying. *BioImpacts, 10*(1), 5–7. https://doi.org/10.15171/bi.2019.30

Mahmoudi, M., & Keashly, L. (2021). Filling the space: A framework for coordinated global actions to Diminish academic bullying. *Angewandte Chemie (International ed. in English), 60*(7), 3338–3344. https://doi.org/10.1002/anie.202009270

Mahmoudi, M., & Moss, S. (2020). The absence of legal remedies following academic bullying. *BioImpacts, 10*(2), 63–64. https://doi.org/10.34172/bi.2020.08

Massey, M., Arif, S., Embuldeniya, S., Nanglu, K., & Bielawski, J. (2022). Ten simple rules for succeeding as an underrepresented STEM undergraduate. *PLoS Computational Biology, 18*(6), e1010101. https://doi.org/10.1371/journal.pcbi.1010101

Miller, H., Seckel, E., White, C. L., Sanchez, D., Rubesova, E., Mueller, C., & Bianco, K. (2022). Gender-based salary differences in academic medicine: a retrospective review of data from six public medical centers in the Western USA. *BMJ Open, 12*(4), e059216. https://doi.org/10.1136/bmjopen-2021-059216

Miller-Kleinhenz, J. M., Kuzmishin Nagy, A. B., Majewska, A. A., Adebayo Michael, A. O., Najmi, S. M., Nguyen, K. H., Van Sciver, R. E., & Fonkoue, I. T. (2021). Let's talk about race: Changing the conversations around race in academia. *Communications Biology, 4*(1), 902. https://doi.org/10.1038/s42003-021-02409-2

Murphy, M., Callander, J. K., Dohan, D., & Grandis, J. R. (2022). Networking practices and gender inequities in academic medicine: Women's and men's perspectives. *EClinicalMedicine, 45*, 101338. https://doi.org/10.1016/j.eclinm.2022.101338

O'Brien, B. C., Artino, A. R., Jr., Costello, J. A., Driessen, E., & Maggio, L. A. (2021). Transparency in peer review: Exploring the content and tone of reviewers' confidential comments to editors. *PLoS One, 16*(11), e0260558. https://doi.org/10.1371/journal.pone.0260558

Pinho-Gomes, A. C., Vassallo, A., Thompson, K., Womersley, K., Norton, R., & Woodward, M. (2021). Representation of women among editors in Chief of Leading Medical Journals. *JAMA Network Open, 4*(9), e2123026. https://doi.org/10.1001/jamanetworkopen.2021.23026

Rohrer, J. M., Tierney, W., Uhlmann, E. L., DeBruine, L. M., Heyman, T., Jones, B., Schmukle, S. C., Silberzahn, R., Willén, R. M., Carlsson, R., Lucas, R. E., Strand, J., Vazire, S., Witt, J. K., Zentall, T. R., Chabris, C. F., & Yarkoni, T. (2021). Putting the self in self-correction: Findings from the loss-of-confidence project. *Perspectives on Psychological Science: A Journal of the Association for Psychological Science, 16*(6), 1255–1269. https://doi.org/10.1177/1745691620964106

Vazire, S., & Holcombe, A. O. (2021). Where are the self-correcting mechanisms in science? *Review of General Psychology, 26*(2), 212–223. https://doi.org/10.1177/10892680211033912

Appendix 5.1

Curriculum to prevent racism in academia

Course learning objective	Lesson sub-objectives
Becoming comfortable having uncomfortable conversations surrounding race and social justice within our own group.	1 Discuss how people may take comfort in their own privilege and how that influences their ability to discuss race, social justice, and action. 2 Define intersectionality, identify examples of intersectionality, and connect this concept to conversations of race. 3 Examine and describe how language and semantics are important in these conversations. 4 Practice engaging in conversations about race.
Deepening our understanding of race as a construct: perceptions on race throughout history and the nature of anti-Black racism in the U.S.	1 Describe African American history, including milestones and lingering systems of anti-Black racism in the United States. 2 Understand the African Diaspora and how the trans-Atlantic slave trade cemented race as a construct in the United States. 3 Understand how BIPOC contributed to the labour movement and how this contribution disproportionately benefited white Americans.
Understanding our personal relationship to white supremacy and anti-Black racism.	1 Define and recognise white supremacy. 2 Gain the understanding that we all have a relationship with white supremacy, and learn to identify the intersectionality of our experiences. 3 Acknowledge and accept past behaviour and learn to move forward.
Evaluating evidence-based practices for mentoring and increasing the retention of BIPOC in STEM.	1 Identify the historical and social constructs that impede the scientific success of BIPOC at various levels of education. 2 Evaluate current programmes that attempt to address these inequities at various levels of education. 3 Brainstorm ways in which to improve programmes that increase retention and representation of BIPOC in STEM at the undergraduate level. 4 Design a framework from which to foster inclusivity in the classroom and pass on (academic) cultural capital to our undergraduate students.
Investigating the impacts of macro- and microaggressions in the classroom and beyond.	1 Easily differentiate between microaggression and macroaggression and become familiar with their different forms or types. 2 Become familiar with the different categories of racial microaggressions. 3 Explain the roots of microaggression and macroaggression. 4 Articulate the impact of repeated (lifetime) macro- and microaggressions on behaviour, growth, and achievement of students and colleagues. 5 Quickly recognise and address microaggressions when you witness or cause one. 6 Appropriately respond to a microaggression and make micro interventions – short-term or transient interventions without a counsellor.

(*Continued*)

(Continued)

Course learning objective	Lesson sub-objectives
Implementing social justice practices in the higher education curriculum.	1 Discuss how recent movements (e.g., #BlackInTheIvory, #BlackAFInSTEM, #PublishingPaidMe) have spurred more awareness about barriers BIPOC scientists and writers face. 2 Evaluate methods for implementing anti-racist practices in our labs, classrooms, departments, and colleges. 3 Brainstorm ways in which to increase transparency in negotiation proceedings, hiring practices, etc. in higher education.
Recognising, alleviating, and dismantling local systems of oppression.	1 Explain the concepts and theories of oppression. 2 Define social justice, economic security, and equality. 3 Identify the role(s) played by human behaviour in the social environment. 4 Explore personal biases and stereotypes that can affect human behaviour. 5 Identify the impact of privilege and oppression and the potential power dynamics of race in the context of how the dynamics of oppression impact the human developmental process. 6 Identify action strategies used to address and dismantle oppression. 7 Identify ways individuals, social movements, and institutions can promote justice and equality and alleviate oppression.
Fostering anti-racist behaviour among our students, academic peers, and community organisations.	1 Define anti-racism. 2 Recognise why we need to take personal responsibility for eliminating racism. 3 Construct your action plan as an anti-racist in your family, department, and classroom. 4 Make a plan on how to continue your journey of becoming anti-racist as an individual and part of an organisation.

Source: Miller-Kleinhenz et al. (2021).

Note: BIPOC, Black, Indigenous, and people of colour.

6   Understanding and addressing questionable research practices

Abstract

The topic of questionable research practices (QRPs) is explored in this chapter. It begins with a definition and description of QRPs, followed by an explanation about how each QRP occurs. The results in the literature about the occurrence of QRPs are then presented along with strategies that can reduce the occurrence of QRPs. Among other stakeholders, it is anticipated that the contents of this chapter will help academic integrity officers identify QRPs and in doing so safeguard the production of research.

Keywords: Cherry Picking; Datasets; HARKing; P-Hacking; Pre-registration; Questionable Research Practices

Key points

- Questionable research practices (QRPs) are predominantly used to adjust the results obtained. Common QRPs are Hypothesising After Results are Known (HARKing), selecting specific pieces of data (i.e., cherry picking), and altering a p-value by selecting a particular statistical tool and/or changing the contents of a dataset (i.e., p-hacking).
- To prevent the occurrence of p-hacking the binary description of 'statistically significant' and 'statistically insignificant' should be replaced with descriptive phrases, such as 'little to no evidence', 'weak evidence', 'moderate evidence', 'strong evidence', and 'very strong evidence'.
- To prevent cherry picking and HARKing studies can be registered and authors should be required to submit to the journal both their dataset and the manuscript for consideration.

6.1   Defining questionable research practices

Banks and colleagues (2016, p. 7) have defined QRPs as:

> design, analytic or reporting practices that have been 'questioned' because of the potential for the practice to be employed with the purpose of presenting biased evidence in favor of an assertion.

QRPs can be conceptualised as being in the middle of a spectrum of most to least ethical research practices. Dubious research practices (e.g., plagiarism, data fabrication, and salami slicing/salami publishing) occupy one end of this spectrum while ethical research practices

DOI: 10.4324/9781003510376-6

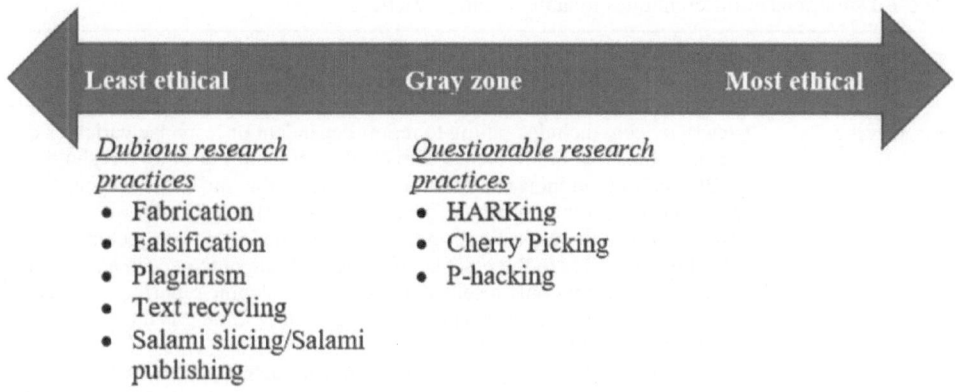

Figure 6.1 Ethical research practices spectrum

occupy the opposite end. In contrast, QRPs occupy a 'grey zone', whereby they are not outright research misconduct nor the most ethical way that research can be conducted (Ravn & Sørensen, 2021) (see Figure 6.1).

There are three common types of QRPs, which are p-hacking, cherry picking, and Hypothesising After Results are Known (HARKing). Each of these QRPs have been defined differently (see Table 6.1).

6.1.1   Cherry picking

The term 'cherry picking' is analogous to fruit pickers who only select fruit that appears palatable and ripe. Researchers who engage in cherry picking only select information that supports pre-existing notions (Andrade, 2021). There are five different types of cherry picking that researchers can either use individually or in combination. First, 'selecting specific records in time' occurs when records that were collected at a specific timeframe are selected for analysis. Second, 'selecting isolated examples' occurs when records that support a pre-formed conclusion are selected. Third, 'selecting responses from specific locations' occurs when data that was collected from one or a few specific locations is selected for analysis. Fourth, 'selecting isolated papers' occurs when studies that support a conclusion are deliberately selected. Fifth, 'quote mining' occurs when a researcher selects quotes that supports their conclusion (Farmer & Cook, 2013).

6.1.2   P-hacking

P-hacking is the act of deliberately changing the *p*-values in a study to either confirm or disprove a hypothesis (Andrade, 2021; Fraser et al., 2018). There are three reasons why researchers engage in p-hacking. First, education institutions place upon them the expectation that they need to publish results that are statistically significant. Second, journals that have high impact factors tend to publish articles that contain statistical significance results that support the hypothesis. Third, agencies that allocate grants base their decisions on the researcher's publication history and the results from previous studies. Thus, to obtain grant funding researchers are incentivised to engage in p-hacking to make their research appealing to prospective funding agencies (Raj et al., 2018).

P-hacking can have three detrimental effects on the production of research. First, studies that contain fraudulent claims of statistical significance, due to the *p*-value being subjected to

Table 6.1 Definitions of different questionable research practices

Questionable research practice	Definition
Cherry picking	"*Cherry picking* includes failing to report dependent or response variables or relationships that did not reach statistical significance or other threshold and/or failing to report conditions or treatments that did not reach statistical significance or other threshold." (Fraser et al., 2018, p. 2). "Selective outcome reporting, or outcome switching or 'cherry-picking' as it is also known, refers to the practice of using multiple outcomes in a research study but reporting only a selection. Selective outcome reporting increases the probability that a statistically significant study finding is due to chance." (Büttner et al., 2020, p. 1366). "The researcher cherry-picks only the significant outcomes for the paper that presents the findings; the nonsignificant outcomes are omitted as though those outcomes had not been studied. Or, when discussing the findings of their study, authors may cherry-pick for consideration research that favors their viewpoint and may criticise or even neglect to cite studies that do not support their arguments. Cherry-picking is a QRP because the reader is deceived into seeing a picture that is more favorable than it truly is." (Andrade, 2021. p. e1).
P-hacking	"*P hacking* refers to a set of activities: checking the statistical significance of results before deciding whether to collect more data; stopping data collection early because results reached statistical significance; deciding whether to exclude data points (e.g., outliers) only after checking the impact on statistical significance and not reporting the impact of the data exclusion; adjusting statistical models, for instance by including or excluding covariates based on the resulting strength of the main effect of interest; and rounding of a p-value to meet a statistical significance threshold (e.g., presenting 0.053 as $P < .05$)." (Fraser et al., 2018, p. 2). "[P-hacking typically] occurs when researchers collect or select data or statistical analyses until nonsignificant results become significant." (Head et al., 2015, p. 1). "*P*-hacking is a QRP wherein a researcher persistently analyzes the data, in different ways, until a statistically significant outcome is obtained; the purpose is not to test a hypothesis but to obtain a significant result. Thus, the researcher may experiment with different statistical approaches to test a hypothesis; or may include or exclude covariates; or may experiment with different cutoff values; or may split groups or combine groups; or may study different subgroups; and the analysis stops either when a significant result is obtained or when the researcher runs out of options. The researcher then reports only the approach that led to the desired result." (Andrade, 2021, p. e2).
Hypothesising After Results are Known (HARKing)	"HARKing includes presenting ad hoc and/or unexpected findings as though they had been predicted all along and presenting exploratory work as though it was confirmatory hypothesis testing." (Fraser et al., 2018, p. 2). "HARKing is defined as presenting a post hoc hypothesis (i.e., one based on or informed by one's results) in one's research report as if it were, in fact, an a priori hypotheses." (Kerr, 1998, p. 196). "HARKing is a QRP wherein a researcher analyses data, observes a (not necessarily expected) statistically significant result, constructs a hypothesis based on that result, *and then presents the result and the hypothesis as though the study had been designed, conducted, and analyzed or at least oriented to test that hypothesis.* The italicised part of the definition constitutes the QRP in HARKing." (Andrade, 2021, p. e1).

a p-hack, can encourage other researchers to conduct similar studies. When this occurs, they inadvertently waste their time and resources pursuing false research leads. Second, studies that contain *p*-values that are caused by p-hacking can be incorporated into systematic reviews and meta-analyses. Their incorporation can significantly change the outcomes of these reviews of the literature. Third, studies intended to improve a person's health that contain incorrect *p*-values and claims of statistical significance can compromise this outcome (Head et al., 2015; Ioannidis, 2005; Raj et al., 2018).

Motulsky (2014) has outlined five distinct ways that a researcher can conduct p-hacking. First, researchers can recruit more participants to increase the sample size. Second, researchers can only examine a specific subset of the dataset that will yield a desired *p*-value. Third, the data in the dataset can be adjusted, perhaps by stratifying the sample. Fourth, outliers and extreme results in the dataset can be removed. Fifth, researchers can use a different statistical test to achieve the desired *p*-value (Motulsky, 2014).

6.1.3   *Hypothesising After Results are Known*

Hypothesising After Results are Known (HARKing) was discovered by social psychologist Norbet Kerr during 1998 (Kerr, 1998). Bishop (2019, p. 435) has characterised HARKing along with p-hacking, publication bias, and low statistical power as 'the four horsemen of the reproducibility apocalypse'. Rubin (2017) has used a shooter analogy to explain the concept of HARKing. In his analogy, a shooter aims and fires several bullets at a target that is painted on a wall. None of the fired bullets hit this target. However, they erase their initial target and draw a revised target around most of the bullet holes to make it appear that they had predominantly hit the target. In this analogy, the researcher is represented by the shooter, the evidence is represented by the bullet holes, the original target represents the priori hypothesis, and the revised target represents the post hoc hypothesis.

Andrade (2021) has given a real-world illustration of HARKing. In their example, a medical researcher formulates the hypothesis that participants who take drug X, an antidepressant, will report that their depressive symptoms are less severe. To test this theory, they divide the participants into two equal groups that are comparable in terms of age, sex, education, socioeconomic status, and baseline depression intensity. One group receives drug X (i.e., experimental group) while the other group receives a placebo treatment (i.e., control group). There is no change in the reported intensity of depression when the course of treatment is completed. As a result, the original hypothesis is incorrect. However, after re-examining the collected data, the medical scientist determines that those in the experimental group who were either overweight or obese had a higher rate of not reacting to the anti-depressant medicine than those who were of normal weight. With this data, the medical scientist develops a new hypothesis: if a person is overweight or obese, the effectiveness of antidepressant treatment is lowered. Based on this new idea, the medical scientist can then write a study. The act of developing a new hypothesis after the findings have been determined (i.e., HARKing) is a QRP because when variables are compared, either across or within groups, one or more will be statistically related. This might lead to a finding (e.g., being fat or overweight can impair the efficacy of antidepressant treatment) being accepted as true when it is likely that the statistically significant connection could have occurred by chance, which is known as a 'Type I statistical Error' (Andrade, 2021).

6.2   Occurrence of questionable research practices

The proportion of the academic community that has witnessed and/or engaged in QRPs has been measured (Artino et al., 2019; Fraser et al., 2018; Gerrits et al., 2019; Gopalakrishna et al., 2022; Hofmann et al., 2020; Hopp & Hoover, 2019; Kaiser et al., 2022; Xie et al., 2021).

Agnoli and colleagues (2017) compared the prevalence of QRPs between psychologists in Italy and the United States of America. They reported that most psychologists in the United States of America failed to report all their study's dependent measures ($n = 486$, 63.4%) while most psychologists in Italy reported that they contemplated if they should collect more data after examining if the results were statistically significant ($n = 222$, 53.2%). In contrast, a minority of psychologists in the United States of America and Italy admitted to falsifying data (i.e., $n = 495$, 0.6% and $n = 220$, 2.3%, respectively) (see Table 6.2).

Table 6.2 Questionable research practices and self-admission rates in percentages for American and Italian psychologists

Questionable research practices	*United States of America*		*Italian association of psychology*	
	Self-admission rate (N)	*95% CI*	*Self-admission rate (N)*	*95% CI*
1 In a paper, failing to report all of a study's dependent measures	63.4 (486)	59.1–67.7	47.9 (219)	41.3–54.6
2 Deciding whether to collect more data after looking to see whether the results were significant	55.9 (490)	51.5–60.3	53.2 (222)	46.6–59.7
3 In a paper, failing to report all of a study's conditions	27.7 (484)	23.7–31.7	16.4 (219)	11.5–21.4
4 Stopping collecting data earlier than planned because one found the result that one had been looking for	15.6 (499)	12.4–18.8	10.4 (221)	6.4–14.4
5 In a paper, "rounding off" a p-value (e.g., reporting that a p-value of 0.054 is less than 0.05)	22.0 (499)	18.4–25.7	22.2 (221)	16.7–27.7
6 In a paper, selectively reporting studies that "worked"	45.8 (485)	41.3–50.2	40.1 (217)	33.6–46.6
7 Deciding whether to exclude data after looking at the impact of doing so on the results	38.2 (484)	33.9–42.6	39.7 (219)	33.3–46.2
8 In a paper, reporting an unexpected ending as having been predicted from the start	27.0 (489)	23.1–30.9	37.4 (219)	31.0–43.9
9 In a paper, claiming that results are unaffected by demographic variables (e.g., gender) when one is actually unsure (or knows that they do)	3.0 (499)	1.5–4.5	3.1 (223)	0.9–5.4
10 Falsifying data	0.6 (495)	0.0–1.3	2.3 (220)	0.3–4.2

Source: Agnoli et al. (2017, p. 4).

Note: Confidence intervals for psychologists in the United States of America were computed from data provided by Leslie John.

6.3   Strategies to reduce questionable research practices

6.3.1   *Using evidence-based language*

Typically, if a p-value is less than the arbitrary 0.05 level the result is typically deemed as 'statistically significant'. Conversely, if it is greater than this arbitrary threshold then the result is not considered statistically significant. However, this binary approach has created a situation in which scientific conclusions and policy decisions are based on if the p-value is above or below the arbitrary p-value threshold of 0.05. To dispel this simplistic approach to interpreting p-values the American Statistical Association (ASA) has released a policy statement of six principles about p-values. Principle three clarifies the importance of p-value thresholds. According to the ASA:

> Scientific conclusions and business or policy decisions should not be based only on whether a p-value passes a specific threshold: Practices that reduce data analysis or scientific inference to mechanical "bright-line" rules (such as "p<0.05") for justifying scientific claims or conclusions can lead to erroneous beliefs and poor decision making. A conclusion does not immediately become "true" on one side of the divide and "false" on the other. Researchers should bring many contextual factors into play to derive scientific inferences, including the design of a study, the quality of the measurements, the external evidence for the phenomenon under study, and the validity of assumptions that underlie the data analysis. Pragmatic considerations often require binary, "yes-no" decisions, but this does not mean that p-values alone can ensure that a decision is correct or incorrect. The widespread use of "statistical significance" (generally interpreted as "p<0.05") as a license for making a claim of a scientific finding (or implied truth) leads to considerable distortion of the scientific process.
>
> (Wasserstein & Lazar, 2016, p. 131)

To dispel the specious and simplistic notion of the binary 'statistically significant' versus 'statistically insignificant' interpretation of p-values, Muff and colleagues (2022) have proposed that evidence-based language should instead be used to describe statistical results.

> Will a seemingly trivial change from a language built around binary statistical significance to a more continuous language of evidence make a real change? We think so, because by reporting results in this way, we automatically move away from drawing unfounded binary conclusions. At the same time, we can free ourselves from hunting arbitrary cut offs that magically determine whether our research was a success or a failure.
>
> (Muff et al., 2022, p. 208).

Muff and colleagues have illustrated a range of evidence-based terms that can be used to describe different p-values (see Figure 6.2). They have also provided four generic examples of how to rewrite results described in terms of statistical significance to evidence-based language (see Table 6.3).

6.3.2   *Justifying specific tests for p-values*

To discover a desired p-value, which is typically one that is either more or less than the established arbitrary 0.05 p-value threshold, some researchers have performed multiple examinations

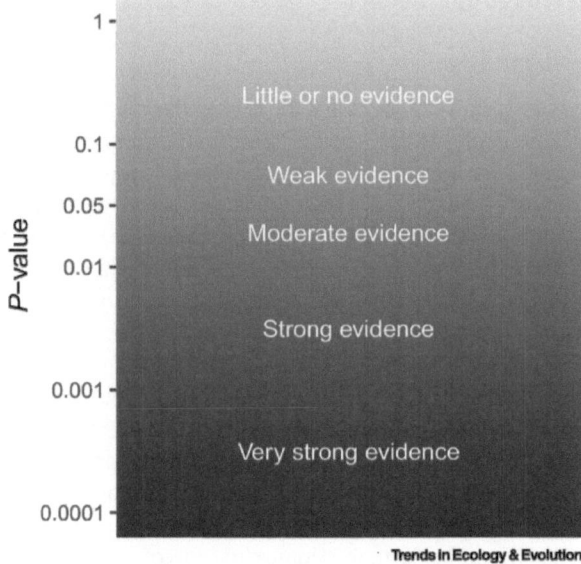

Figure 6.2 Suggested ranges to approximately translate the *p*-value into the language of evidence

Source: Muff et al. (2022, p. 206).

Note: Boundaries should not be understood as absolute thresholds.

Table 6.3 Generic examples of how to rewrite results from the statistical significance to evidence-based language

Statistical significance language	Evidence language (suggestions)
The effect of *x* on *y* was not significant (*p* = 0.53).	There was no evidence that *x* has an effect on *y* [(give effect estimate), *p* = 0.53]. The data did not have any evidence about the direction of any association of *x* with *y* [(give effect estimate), *p* = 0.53].
The effect of *x* on *y* was not significant (*p* = 0.08).	There was (only) weak evidence that *x* (positively/negatively) affects *y* [(give effect estimate), *p* = 0.08]. There was (only) weak evidence that *x* is (positively/negatively) associated with *y* [(give effect estimate), *p* = 0.08].
The effect of *x* on *y* was significant (*p* = 0.03).	There was (only) moderate evidence that *x* has a (positive/negative) effect on *y* [(give effect estimate), *p* = 0.03]. The data revealed moderate evidence that *x* is (positively/negatively) associated with *y* [(give effect estimate), *p* = 0.03].
The effect of *x* on *y* was significant (*p* = 0.0003).	There was very strong evidence for a (positive/negative) effect of *x* on *y* [(give effect estimate), *p* < 0.001]. The data revealed very strong evidence that *x* is (positively/negatively) associated with *y* [(give effect estimate), *p* < 0.001].

Source: Muff et al. (2022, p. 207).

Note: A general recommendation is that *p*-values should be accompanied by effect size estimates whenever possible. The remark whether an effect was positive or negative should be added when this is possible (e.g. for continuous variables *x*)

of the dataset. To help prevent this practice from occurring it has been proposed that researchers should justify all decisions about why they used specific statistical tests (Betensky, 2019). The ASA has also supported this view, stating that:

> Proper inference requires full reporting and transparency P-values and related analyses should not be reported selectively. Conducting multiple analyses of the data and reporting only those with certain p-values (typically those passing a significance threshold) renders the reported p-values essentially uninterpretable. Cherry picking promising findings, also known by such terms as data dredging, significance chasing, significance questing, selective inference, and "p-hacking," leads to a spurious excess of statistically significant results in the published literature and should be vigorously avoided. One need not formally carry out multiple statistical tests for this problem to arise: Whenever a researcher chooses what to present based on statistical results, valid interpretation of those results is severely compromised if the reader is not informed of the choice and its basis. Researchers should disclose the number of hypotheses explored during the study, all data collection decisions, all statistical analyses conducted, and all p-values computed. Valid scientific conclusions based on p-values and related statistics cannot be drawn without at least knowing how many and which analyses were conducted, and how those analyses (including p-values) were selected for reporting.
>
> (Wasserstein & Lazar, 2016, pp. 131–132)

6.3.3   *Pre-registering a study's design*

The *Center for Open Science* has explained the steps for the pre-registration of studies. They explained that a researcher begins with a research proposal that includes details about the study's research question, sample size, recruitment strategies, data collection method (e.g., surveys, interviews, or mixed-methods), recruitment criteria, strategies to examine the collected data, data retention policies, ethical concerns, sources of funding, etc. (Rauh et al., 2020; Wright et al., 2020). The research proposal is then submitted for peer review. This review ensures that it is of sufficient quality to both achieve the objective of the study and be reproducible. If the research proposal is approved, then the researcher receives a guarantee that their study will be published regardless of its results. This guarantee can ensure that publication bias is mitigated. The researcher then conducts the study in accordance with the approved research proposal. Afterwards, they submit their manuscript for peer review. If approved after peer review it is published. Alternatively, if the manuscript is inadequate then it is either rejected or published after it has been re-drafted (see Figure 6.3).

To prevent researchers from engaging in QRPs, particularly p-hacking and HARKing, journals can request that the author first provides a pre-registration of the study. To prevent the researcher from using HARKing to make their study appealing their pre-registered research proposal should state the hypothesis or hypotheses that they will test. In reference to p-hacking, once the data is collected a researcher can use at least nine different techniques to change a p-value (Motulsky, 2014). To prevent them from doing such acts the research proposal should contain details about the study's sample size, statistical tests that will be used to examine the data, and assurances that all the collected data will be published along with the study.

Figure 6.3 The Center for Open Science outline for pre-registration of studies
Source: Center for Open Science (2021).

6.3.4 *Reforming grant awarding agencies*

Diong and colleagues (2021) evaluated the funding criteria of eight health and medical funding agencies in Australia. They concluded that most agencies did not require:

- publicly registering study protocols before starting data collection,
- publicly registering analysis protocols before starting data analysis, and
- the researcher to consult with a statistician.

In Australia, and around the world, agencies that allocate funding for research should adhere more to the principles that foster the creation of reproducible research (Rauh et al., 2020; Wright et al., 2020). Additionally, these agencies should also adhere to all Diong et al.'s criteria for the responsible conduct of research. Failure to implement such guidelines has the potential to increase the prospect of QRPs occurring.

6.3.5 *Educating scholars about questionable research practices*

Scholars who lack education about QRPs can inadvertently incorporate into their research such practices. Olsen and colleagues (2019) examined the proportion of QRPs among 250 psychology students' master's theses. They reported that while no obvious indications of p-hacking occurred, they discovered that in 18% of cases inconsistent *p*-values were documented. To reduce this figure, researchers and early career researchers should receive from their academic institution education about QRPs. Such training can reduce the prospect of them engaging in QRPs since after receiving such training they will not be able to use ignorance to justify using QRPs. To ensure that researchers have understood the topic of QRPs such training courses can use Fraser et al.'s (2018) survey about QRPs as a guide to measure this understanding (see Appendix 6.1).

6.3.6 *Creating reporting procedures*

Academic seniority and the precariat nature of academic employment can prevent academic misconduct and fraud from being reported. Horbach and colleagues (2020) explained that junior researchers rarely reported senior academics engaging in academic misconduct because they are concerned that it will have a detrimental impact on their ability to complete their research and subsequently have an academic career. Similarly, junior academics are more inclined to be employed on a temporary basis and consequently would be less likely to report academic misconduct out of fear of not being continually employed. Based on the collected insights Horbach and colleagues (2020, p. 1615) proposed that:

> Our results suggest a need for improved reporting procedures. Specifically, such procedures should take the position of an organisation's less powerful members, such as junior researchers and people with temporary work appointments, into account and facilitate their reporting. This requires procedures that effectively address issues of power imbalance and the fear of not being taken seriously. The implementation of such procedures could help target a culture of complacency and cynicism that normalises questionable research practices. In addition, it may contribute to a sense of organisational responsibility that should ultimately foster a climate of research integrity.
>
> (Horbach et al., 2020, p. 1615)

Universities and other research institutions should either establish or modify their reporting procedures using Horbach and colleagues' proposition as a guide. Among other actions, universities should make accommodations to protect junior academic staff who are temporarily employed against retribution for reporting academic misconduct. Universities and research institutions should also implement education campaigns and other strategies to purge any culture of complacency and cynicism within their research environments.

6.3.7   *Reforming the 'publish or perish' culture*

> The current evaluation criteria are thought to create a perverse incentive structure, are unidimensionally focused on the "bean counting" of publications in high-impact and branded journals, and may nudge researchers unconsciously into QRP with a focus on publishing as many articles as possible, instead of getting it right. How these incentive structures actually influence research practices is still unknown and how the research evaluation criteria should be reformed needs further research.
>
> (Tijdink et al., 2021, p. 456)

As described above, Tijdink and colleagues (2021) explained that within tertiary education institutions newly minted academics are conditioned to publish their manuscripts in journals that have high impact factors. Often the competition to publish within such journals can be intensive. When confronted with this competition, along with limited experiences and skills in creating high-quality manuscripts, some inexperienced academics might resort to using QRPs. The rationale for using QRPs is to create a manuscript that is sensationalist and enticing for journals with high impact factors.

Two modifications to recruiting and employing inexperienced academics need to occur to dissuade them from using QRPs to maximise the prospect of their study being published. First, the criteria used to employ academics should not place a large emphasis on the number of manuscripts that they have published or the prestige where their manuscripts were published, as denoted by the journal's impact factor. Instead, the criteria used to determine who should be appointed to an academic position should focus on the candidate's knowledge and proficiency about ethical research practices. To demonstrate this understanding candidates should explain what are QRPs and what they have done or will do if they discover a colleague in the academic workplace engaging in QRPs or other academic misconduct. Second, universities need to provide training to inexperienced researchers about QRPs and inform them about the professional consequences that using QRPs can have on their careers. This training can legally protect universities against accusations of unfair dismissal from terminated employees who were found to have engaged in QRPs.

6.3.8   *Removing any financial incentives for academic publishing*

> There has been recent interest in the revelation that in China cash payments are made to authors who publish in internationally rated journals. These payments range from $1000 for publication in PLOS 1 to $50,000, in Nature or Science. I would ask readers of this article to think about whether their behaviour might be influenced by large sums of money. Is there a sum which would be sufficient? While it may be that the offer of $50,000 would not perturb your moral compass, would a billion be enough? It also would not be fair to focus on China. A recent letter to Science points out that such bonuses are seen in many countries including the USA & UK.
>
> (Eisner, 2018, p. 365)

Eisner (2018), in the quotation above, explained that paying academics for manuscript publication places some into a position where they are motivated to publish for money instead of a desire to advance knowledge. When confronted with such financial incentives some academics might be tempted to use QRPs to improve their chances of their manuscript being published so that they can receive a financial reward. This temptation is further exacerbated when the manuscript's author lacks financial security because they are employed in a temporary academic position. To prevent academics from using QRPs to improve their chances of publishing manuscripts for financial gain universities and research institutes must not financially reward academics for their publications.

6.3.9   *Creating an independent research integrity agency*

> … consider the way that it [fraud] is investigated. The responsibility for deciding whether fraud has occurred rests, not with the journal, but, rather, with the institution, and in some cases the funder. I find it hard to think that nobody else can see the obvious conflict of interest. The person under investigation may be someone with enormous grant income who brings great prestige to the institution. It may therefore not always be in the best interest of the institution to investigate too thoroughly.
>
> (Eisner, 2018, p. 365)

Above Eisner (2018) explains that for some research institutions investigating accusations of academic misconduct is a conflict of interest because to do so would undermine their prestige and ability to attract research grants. To prevent academic misconduct from occurring, Grey and colleagues (2019) have proposed that an external agency responsible for investigating allegations of academic misconduct and fraud should be established. Such an agency could be a viable option for research staff who believe that their employer's internal policies and practices for investigating research misconduct are insufficient (Grey et al., 2019). This sentiment was echoed by Grey and colleagues (2019, p. 5), who once stated:

> Our analyses identify important deficiencies in the quality and reporting of institutional investigation of concerns about research integrity. They reinforce disquiet about the ability of institutions to rigorously and objectively oversee the integrity of research conducted by their

own employees and the lack of regulatory oversight. A possible solution is the establishment of more efficient and adequately resourced independent organisations with the authority and expertise to undertake and report investigations, and implement recommendations, including those which span multiple institutions and countries.

An independent office or agency, funded by the government, that is tasked with monitoring research integrity and QRP usage is not a novel concept. Such agencies already exist in many developed nations, such as the United States of America (i.e., The Office of Research Integrity), the United Kingdom (i.e., UK Research Integrity Office), and Sweden (i.e., Research Misconduct Board). However, despite calls for such an agency occurring since 2018, such an agency does not exist within Australia (Vaux, 2022; Vaux et al., 2018). This is perhaps somewhat surprising considering that Australia is an industrialised nation and that the Australian government spends billions of dollars on research endeavours that is distributed via the *National Health and Medical Research Council* and the *Australian Research Council*. To preserve the intent of this funding, and the public's trust in research, such an agency should be established as soon as feasible.

6.3.10   *Making researchers pledge an oath to uphold research integrity*

To deter researchers from engaging in research misconduct within France all PhD candidates are required to pledge an oath to research ethics and integrity. A draft of this oath, which was published by *Science*, reads in part: 'I pledge, to the greatest of my ability, to continue to maintain integrity in my relationship to knowledge, to my methods and to my results' (Rabesandratana, 2022, p. 251). This oath will be affirmed by all PhD candidates at the start and end of their doctoral education. Hugh Desmond claimed that such an oath would 'strengthen a sense of professionalism among researchers, help coordinate norms, and make them public. [It could] empower researchers that are lower in the hierarchy, and liberate more senior researchers' (Rabesandratana, 2022, p. 251). However, although a symbolic and well-intentioned gesture, many believe that such an oath will not protect or improve the practice of research integrity. For example, Josefin Sundin claimed that 'The only way to improve research integrity is to promote and reward research rigor, transparency, and reproducibility over impact factor and number of publications' (Rabesandratana, 2022, p. 251). Perhaps to help abate any concerns or criticisms such an oath could be enforced by rules that ensure that perpetrators of research misconduct receive the appropriate consequences.

6.3.11   *Developing a confidential reporting system*

To help stop the usage of QRPs, research institutions, universities, and colleges can implement an anonymous system for reporting such allegations of academic misconduct (Fischhoff et al., 2021). Fischhoff and colleagues (2021) have proposed a four-stage process of lodging and investigating allegations of academic misconduct and QRPs, which are individual submission, report monitoring, report coordination, and action (see Table 6.4).

6.3.12   *Aubert Bonn and colleagues' suggestions about improving research integrity*

Aubert Bonn and colleagues (2022) collected comments from a diverse range of stakeholders over the last five years to better understand the impact that existing methods of creating research

Table 6.4 Four-stage process of lodging and investigating allegations of academic misconduct and fraud proposed by Fischhoff and colleagues

Action	Description/excerpt
Individual submission	Early career scientists who have observed apparent misconduct file a confidential report with their institution's *scientific integrity official*. That official helps them to interpret their experience, provides emotional support, and describes their options. If they choose to proceed on their own, they will be referred to the proper channels. If they wish to proceed, *but not alone*, they can file a report that remains entirely under their control. If they choose to end the inquiry, all records are erased.
Report monitoring	The scientific integrity official monitors reports, looking for patterns of abuse associated with particular scientists or labs. If a pattern emerges, the official contacts each person who has filed a relevant report. The official describes the existence of other reports but says nothing about their contents or sources. The official asks filers which elements of their reports, if any, can be shared with other filers.
Report coordination	The official shares information consistent with filers' instructions. That sharing might entail anything from revealing no report details to convening a joint meeting. The official documents the process, including when sharing has breached the independence of the evidence. The official informs those who have not joined the process about its progress, while protecting the confidentiality of those who have joined.
Action	Should some, or all, of the filers choose to proceed, the official informs them of their options and facilitates transferring the case to the proper jurisdiction. In order to ensure fairness to all parties, including the right to self-defence against charges, no action can be taken when senior scientists are deceased or incapacitated.
	This procedure seeks to make joint action feasible, while preserving filers' control over their own involvement. It may lead to *acceptable inaction*, if filers come to see the apparent misconduct more charitably. It might lead to *bitter inaction*, if filers conclude that they cannot take the risk of pursuing legitimate concerns. It might lead to *joint action*, if a safe forum exists. It might lead to *individual action*, if a filer is ready to risk having to leave the field.

Source: Fischhoff et al. (2021).

may have on the integrity and quality of research. They catalogued the data they collected into four initiatives to promote better research, which were: (1) restructuring the organisation of research, (2) realigning research assessments, (3) re-modelling research careers, and (4) recognising and coordinating efforts to move from reaction to action (see Figure 6.4) (Aubert Bonn et al., 2022). What follows now is a description of these themes.

6.3.12.1 *Restructure the organisation of research*

While factors inside research environments, such as rivalry, pressure, and incentives, have been found to disrupt research integrity, efforts to enhance integrity often focus on researchers' awareness and compliance rather than the systemic challenges of academia. Aubert Bonn et al.'s findings demonstrate that the promotion of research integrity necessitates interventions that address systemic flaws in the research enterprise including, most importantly, the incentives and reward structures of research. This is not to diminish the importance of integrity training, codes of conduct, whistle-blower protection, and oversight in creating a strong culture of integrity among researchers (Aubert Bonn et al., 2022).

1. **RESTRUCTURE THE ORGANIZATION OF RESEARCH**

Foster research integrity by addressing systemic problems embedded in the core of research structures

Pressures   Competition

Demands   Reward systems

2. **REALIGN RESEARCH ASSESSMENTS**

Realign all levels of assessments to recognise the quality of research and foster research integrity

Openness   Transparency

Quality   Creativity

From
REACTION
to
ACTION

3. **DIVERSIFY AND SECURE RESEARCH CAREERS**

Bridge the gap between academic and non-academic careers and value greater diversity within academia

Employment continuity
Team and collaborative work
Diversity of skills and profiles

4. **RECOGNIZE DISPARATE VOICES AND COORDINATE ACTIONS**

Promote a shared discussion in which all stakeholders can take part to build a strategy for change

Hear all voices
Ensure a proactive dialogue
Join forces and coordinate actions

Figure 6.4 Four recommendations to help overcome the current problems in research creation
Source: Aubert Bonn et al. (2022, p. 3).

6.3.12.2   Realign research assessments

Aubert Bonn et al.'s findings, which echo statements from the San Francisco Declaration on Research Assessment and the Hong Kong Principles for Assessing Researchers, show that research assessments must change in a way that values and fosters integrity and quality. For example, they discovered that researchers felt that the metrics currently used to judge research careers do not correlate with the indicators that are vital for advancing science (Aubert Bonn et al., 2022). A focus on outputs, quantity, and ground-breaking findings can inhibit the production of high-quality research and fail to recognise the significance of negative results and the importance of replication in research. This problem is exacerbated by the increase in project-based research financing, which puts more pressure on researchers and views research as having a short-term focus. Aubert Bonn et al.'s results offer evidence to support the significant impetus for change that is evident in continuing campaigns to promote more responsible use of metrics, wider consideration of various research activities, and increased acknowledgement of procedures like quality, openness, and transparency. However, to realign research assessments, other actions are needed beyond alterations to research institutions. Furthermore, national and worldwide university rankings and performance-based research funding have a tremendous impact on perceptions of success and excellence, and realigning these high-level evaluations with honesty and high-quality research is equally crucial in advancing research (Aubert Bonn et al., 2022).

6.3.12.3 *Diversify and secure research careers*

Aubert Bonn and colleagues found that it is necessary to consider the individual doing the study. Currently, approximately 10–20% of PhD candidates will be able to acquire a permanent post in academia, despite most desiring an academic career. This is not a new discovery since this topic has been highlighted for more than two decades with little progress. The scarcity of employment prospects also increases pressure and rivalry among early career researchers, isolating them, risking their mental health, and forcing them to outperform their peers so that they can survive in their academic career. To stay ahead of their peers, researchers must focus on outputs and disregard processes that are not rewarded, even though many of these processes are critical to scientific advancement. Highly selective research occupations also stifle diversity, not just in terms of gender and race but also in terms of the talents and career paths of those who succeed. As a result, academic research environments are influenced by a consistent resistance to changing research culture. There is a pressing need to address research careers and job instability. Research institutes and doctorate schools must provide greater possibilities for early career researchers to build transferrable skills and engage with non-academic industries. However, academic professions would benefit from increased differentiation as well, including multiple jobs within academia where distinct abilities and profiles are recognised, motivated, and rewarded, as well as collaborative teams and diverse interpretations of success (Aubert Bonn et al., 2022).

6.3.12.4 *Recognise disparate voices and coordinate actions*

Policymakers, research funders, research institution heads, editors and publishers, research integrity office members, early-, mid-, and late-career academics, research students, laboratory staff, and researchers who left academia were all involved in Aubert Bonn et al.'s study. Hearing the opinions of so many various stakeholders, they found that individuals' perceptions on success, integrity, and wrongdoing varied, and that the issues and actions required are viewed differently by different stakeholders. They also discovered that responsibility for acts is frequently transmitted from one player to the next, resulting in a stagnant system marked by blame, despondency, and inaction. Despite this depressing conclusion, experts motivated to change the status quo have emerged in recent years. The San Francisco Declaration on Research Assessments, is an example of researchers acting together to capture the attention of important funders, including the Wellcome Trust in the United Kingdom and the Australian Government's National Health and Medical Research Council. The discussion is also becoming more diversified, bringing together the views of many stakeholders eager to work together to improve research. However, the opinions of former researchers and early career scientists, whose viewpoints may differ significantly from those of individuals who survived and excelled in the existing system, are frequently overlooked. To operationalise large, systemic changes, understanding the dynamics and linkages at work in existing problems as perceived by all parties involved is essential. Broad expert groups, such as the European Commission Policy Platforms or expert groups formed by Scientific Societies and Academies, provide a forum for diverse actors' perspectives to congregate and impact those who shape scientific and research policy. The next natural step in maintaining a proactive discourse is to ensure that these forums feature a diverse range of viewpoints (Aubert Bonn et al., 2022).

6.3.12.5   *From reaction to action*

To improve research integrity, the basic underpinnings of research systems must be addressed. Despite the challenges, Aubert Bonn and colleagues are confident that change is achievable. Much has transpired in the field of research integrity. New advances, evaluation projects, position statements, and prominent opinions on the matter emerged virtually every week while Aubert Bonn and colleagues were conducting their study. Nonetheless, their study indicates that efforts to improve research integrity can only reach their full potential and influence research culture if they produce wide and coordinated change methods. Recent steps in this direction are encouraging. The recent 'Agreement on Reforming Research Assessment', supported by the European Commission, Science Europe, and the European University Association, the Global Research Council's 'Responsible Research Assessment – Call to Action', and statements from broad-reaching multi-stakeholder programmes such as the 'G7 2021 Research Compact' and the 'UNESCO Recommendation on Open Science' all suggest global change (Aubert Bonn et al., 2022).

6.4   Conclusion

The purpose of this chapter was to explain some common QRPs that are used to maximise the prospect of research being published. To understand this purpose a definition and description of three common QRPs were presented. This was followed by an explanation about the occurrence of QRPs in tertiary education and research sectors. A series of recommendations about how research practices can change to inhibit the usage of QRPs concluded this chapter. The contribution that this chapter makes to the research community is to improve the integrity of academic research by reducing the prospect of QRPs occurring.

Additional readings

Cairns, A. C., Linville, C., Garcia, T., Bridges, B., Tanona, S., Herington, J., & Laverty, J. T. (2021). A phenomenographic study of scientists' beliefs about the causes of scientists' research misconduct. *Research Ethics, 17*(4), 501–521. https://doi.org/10.1177/17470161211042658

Kalichman, M. (2020). Survey study of research integrity officers' perceptions of research practices associated with instances of research misconduct. *Research Integrity and Peer Review, 5*(1), 17. https://doi.org/10.1186/s41073-020-00103-1

Krishna, A., & Peter, S. M. (2018). Questionable research practices in student final theses – Prevalence, attitudes, and the role of the supervisor's perceived attitudes. *PLoS One, 13*(8), e0203470. https://doi.org/10.1371/journal.pone.0203470

Maggio, L., Dong, T., Driessen, E., & Artino, A., Jr (2019). Factors associated with scientific misconduct and questionable research practices in health professions education. *Perspectives on Medical Education, 8*(2), 74–82. https://doi.org/10.1007/s40037-019-0501-x

References

Agnoli, F., Wicherts, J. M., Veldkamp, C. L., Albiero, P., & Cubelli, R. (2017). Questionable research practices among italian research psychologists. *PLoS One, 12*(3), e0172792. https://doi.org/10.1371/journal.pone.0172792

Andrade, C. (2021). HARKing, Cherry-picking, P-hacking, Fishing expeditions, and data dredging and mining as questionable research practices. *The Journal of Clinical Psychiatry, 82*(1), 20f13804. https://doi.org/10.4088/JCP.20f13804

Artino, A. R., Jr., Driessen, E. W., & Maggio, L. A. (2019). Ethical shades of gray: International frequency of scientific misconduct and questionable research practices in health professions education. *Academic Medicine: Journal of the Association of American Medical Colleges, 94*(1), 76–84. https://doi.org/10.1097/ACM.0000000000002412

Aubert Bonn, N., De Vries, R. G., & Pinxten, W. (2022). The failure of success: Four lessons learned in five years of research on research integrity and research assessments. *BMC Research Notes, 15*(1), 309. https://doi.org/10.1186/s13104-022-06191-0

Banks, G. C., O'Boyle, E. H., Pollack, J. M., White, C. D., Batchelor, J. H., Whelpley, C. E., Abston, K. A., Bennett, A. A., & Adkins, C. L. (2016). Questions about questionable research practices in the field of management: A guest commentary. *Journal of Management, 41*(1), 5–20. https://doi.org/10.1177/0149206315619011

Betensky, R. A. (2019). The p-value requires context, not a threshold. *The American Statistician, 73*(suppl 1), 115–117. https://doi.org/10.1080/00031305.2018.1529624

Bishop, D. (2019). Rein in the four horsemen of irreproducibility. *Nature, 568*(7753), 435. https://doi.org/10.1038/d41586-019-01307-2

Büttner, F., Toomey, E., McClean, S., Roe, M., & Delahunt, E. (2020). Are questionable research practices facilitating new discoveries in sport and exercise medicine? The proportion of supported hypotheses is implausibly high. *British Journal of Sports Medicine, 54*(22), 1365–1371. https://doi.org/10.1136/bjsports-2019-101863

Center for Open Science. (2021, November 29). *Simple Registered Report Protocol Preregistration.* https://osf.io/rr/

Diong, J., Kroeger, C. M., Reynolds, K. J., Barnett, A., & Bero, L. A. (2021). Strengthening the incentives for responsible research practices in Australian health and medical research funding. *Research Integrity and Peer Review, 6*(1), 11. https://doi.org/10.1186/s41073-021-00113-7

Eisner, D. A. (2018). Reproducibility of science: Fraud, impact factors and carelessness. *Journal of Molecular and Cellular Cardiology, 114*, 364–368. https://doi.org/10.1016/j.yjmcc.2017.10.009

Farmer, G. T., & Cook, J. (2013). Understanding climate change denial. In *Climate Change Science: A Modern Synthesis* (pp. 445–466). Springer, Dordrecht. https://doi.org/10.1007/978-94-007-5757-8_23

Fischhoff, B., Dewitt, B., Sahlin, N. E., & Davis, A. (2021). A secure procedure for early career scientists to report apparent misconduct. *Life Sciences, Society and Policy, 17*(1), 2. https://doi.org/10.1186/s40504-020-00110-6

Fraser, H., Parker, T., Nakagawa, S., Barnett, A., & Fidler, F. (2018). Questionable research practices in ecology and evolution. *PLoS One, 13*(7), e0200303. https://doi.org/10.1371/journal.pone.0200303

Gerrits, R. G., Jansen, T., Mulyanto, J., van den Berg, M. J., Klazinga, N. S., & Kringos, D. S. (2019). Occurrence and nature of questionable research practices in the reporting of messages and conclusions in international scientific Health Services Research publications: A structured assessment of publications authored by researchers in the Netherlands. *BMJ Open, 9*(5), e027903. https://doi.org/10.1136/bmjopen-2018-027903

Gopalakrishna, G., Ter Riet, G., Vink, G., Stoop, I., Wicherts, J. M., & Bouter, L. M. (2022). Prevalence of questionable research practices, research misconduct and their potential explanatory factors: A survey among academic researchers in The Netherlands. *PLoS One, 17*(2), e0263023. https://doi.org/10.1371/journal.pone.0263023

Grey, A., Bolland, M., Gamble, G., & Avenell, A. (2019). Quality of reports of investigations of research integrity by academic institutions. *Research Integrity and Peer Review, 4*, 3. https://doi.org/10.1186/s41073-019-0062-x

Head, M. L., Holman, L., Lanfear, R., Kahn, A. T., & Jennions, M. D. (2015). The extent and consequences of p-hacking in science. *PLoS Biology, 13*(3), e1002106. https://doi.org/10.1371/journal.pbio.1002106

Hofmann, B., Bredahl Jensen, L., Eriksen, M. B., Helgesson, G., Juth, N., & Holm, S. (2020). Research integrity among PhD students at the faculty of medicine: A comparison of three Scandinavian

Universities. *Journal of Empirical Research on Human Research Ethics, 15*(4), 320–329. https://doi.org/10.1177/1556264620929230

Hopp, C., & Hoover, G. A. (2019). What crisis? Management researchers' experiences with and views of scholarly misconduct. *Science and Engineering Ethics, 25*(5), 1549–1588. https://doi.org/10.1007/s11948-018-0079-4

Horbach, S. P., Breit, E., Halffman, W., & Mamelund, S. E. (2020). On the willingness to report and the consequences of reporting research misconduct: The role of power relations. *Science and Engineering Ethics, 26*(3), 1595–1623. https://doi.org/10.1007/s11948-020-00202-8

Ioannidis, J. P. (2005). Why most published research findings are false. *PLoS Medicine, 2*(8), e124. https://doi.org/10.1371/journal.pmed.0020124

Kaiser, M., Drivdal, L., Hjellbrekke, J., Ingierd, H., & Rekdal, O. B. (2022). Questionable research practices and misconduct among Norwegian researchers. *Science and Engineering Ethics, 28*(1), 2. https://doi.org/10.1007/s11948-021-00351-4

Kerr, N. L. (1998). HARKing: Hypothesizing after the results are known. *Personality and Social Psychology Review, 2*(3), 196–217. https://doi.org/10.1207/s15327957pspr0203_4

Motulsky, H. J. (2014). Common misconceptions about data analysis and statistics. *Pharmacology Research & Perspectives, 3*(1), e00093. https://doi.org/10.1002/prp2.93

Muff, S., Nilsen, E. B., O'Hara, R. B., & Nater, C. R. (2022). Rewriting results sections in the language of evidence. *Trends in Ecology & Evolution, 37*(3), 203–210. https://doi.org/10.1016/j.tree.2021.10.009

Olsen, J., Mosen, J., Voracek, M., & Kirchler, E. (2019). Research practices and statistical reporting quality in 250 economic psychology master's theses: A meta-research investigation. *Royal Society Open Science, 6*(12), 190738. https://doi.org/10.1098/rsos.190738

Rabesandratana, T. (2022). France introduces research integrity oath. *Science (New York, N.Y.), 377*(6603), 251. https://doi.org/10.1126/science.add9092

Raj, A. T., Patil, S., Sarode, S., & Salameh, Z. (2018). P-hacking: A wake-up call for the scientific community. *Science and Engineering Ethics, 24*(6), 1813–1814. https://doi.org/10.1007/s11948-017-9984-1

Rauh, S., Torgerson, T., Johnson, A. L., Pollard, J., Tritz, D., & Vassar, M. (2020). Reproducible and transparent research practices in published neurology research. *Research Integrity and Peer Review, 5*, 5. https://doi.org/10.1186/s41073-020-0091-5

Ravn, T., & Sørensen, M. P. (2021). Exploring the gray area: Similarities and differences in questionable research practices (QRPs) across main areas of research. *Science and Engineering Ethics, 27*(4), 40. https://doi.org/10.1007/s11948-021-00310-z

Rubin, M. (2017). When does HARKing hurt? Identifying when different types of undisclosed post hoc hypothesizing harm scientific progress. *Review of General Psychology, 21*(4), 308–320. https://doi.org/10.1037/gpr0000128

Tijdink, J. K., Horbach, S., Nuijten, M. B., & O'Neill, G. (2021). Towards a research agenda for promoting responsible research practices. *Journal of Empirical Research on Human Research Ethics, 16*(4), 450–460. https://doi.org/10.1177/15562646211018916

Vaux, D. (2022). *Australia needs an Office for Research Integrity to catch up with the rest of the world.* The Conversation. https://theconversation.com/australia-needs-an-office-for-research-integrity-to-catch-up-with-the-rest-of-the-world-176019

Vaux, D., Brooks, P., & Gandevia, S. (2018). *Weakened code risks Australia's reputation for research integrity.* The Conversation. https://theconversation.com/weakened-code-risks-australias-reputation-for-research-integrity-98622

Wasserstein, R. L., & Lazar, N. A. (2016). The ASA statement on p-values: Context, process, and purpose. *The American Statistician, 70*(2), 129–133. https://doi.org/10.1080/00031305.2016.1154108

Wright, B. D., Vo, N., Nolan, J., Johnson, A. L., Braaten, T., Tritz, D., & Vassar, M. (2020). An analysis of key indicators of reproducibility in radiology. *Insights into Imaging, 11*(1), 65. https://doi.org/10.1186/s13244-020-00870-x

Xie, Y., Wang, K., & Kong, Y. (2021). Prevalence of research misconduct and questionable research practices: A systematic review and meta-analysis. *Science and Engineering Ethics, 27*(4), 41. https://doi.org/10.1007/s11948-021-00314-9

Appendix 6.1

Survey to measure questionable research practices

Not reporting studies or variables that failed to reach statistical significance (e.g., $p \leq 0.05$) or some other desired statistical threshold.

a) *Please estimate the percentage of researchers who you believe have engaged in this practice on at least one occasion:*

   0   10   20   30   40   50   60 70   80   90   100

b) *Have you ever engaged in this practice?*

   - Never □
   - Once □
   - Occasionally □
   - Frequently □
   - Almost always □

c) *What's your option of this practice?*

   - It should be used almost always □
   - It should be used often □
   - It should only be used rarely □
   - It should never be used □

d) *Optional Question: Why do you think this practice should or shouldn't be used?*

   _____
   _____
   _____

Not reporting covariates that failed to reach statistical significance (e.g., $p \leq 0.05$) or some other desired statistical threshold.

a) *Please estimate the percentage of researchers who you believe have engaged in this practice on at least one occasion:*

   0   10   20   30   40   50   60 70   80   90   100

b) *Have you ever engaged in this practice?*

   - Never □
   - Once □

- Occasionally   □
- Frequently   □
- Almost always   □

c) *What's your option of this practice?*

- It should be used almost always   □
- It should be used often   □
- It should only be used rarely   □
- It should never be used   □

d) *Optional Question: Why do you think this practice should or shouldn't be used?*

_____

_____

_____

Not Reporting an unexpected finding or a result from exploratory analysis as having been predicted from the start.

a) *Please estimate the percentage of researchers who you believe have engaged in this practice on at least one occasion:*

0   10   20   30   40   50   60   70   80   90   100

b) *Have you ever engaged in this practice?*

- Never   □
- Once   □
- Occasionally   □
- Frequently   □
- Almost always   □

c) *What's your option of this practice?*

- It should be used almost always   □
- It should be used often   □
- It should only be used rarely   □
- It should never be used   □

d) *Optional Question: Why do you think this practice should or shouldn't be used?*

_____

_____

_____

Reporting a set of statistical models as the complete tested set when other candidate models were also tested.

a) *Please estimate the percentage of researchers who you believe have engaged in this practice on at least one occasion:*

0   10   20   30   40   50   60   70   80   90   100

b) *Have you ever engaged in this practice?*

- Never   □
- Once   □
- Occasionally   □
- Frequently   □
- Almost always   □

c) *What's your option of this practice?*

- It should be used almost always   □
- It should be used often   □
- It should only be used rarely   □
- It should never be used   □

d) *Optional Question: Why do you think this practice should or shouldn't be used?*

_____

_____

_____

Rounding-off a p-value or other quantity to meet a pre-specified threshold (e.g., reporting $p = 0.054$ as $p = 0.05$ or $p = 0.013$ as $p = 0.01$).

a) *Please estimate the percentage of researchers who you believe have engaged in this practice on at least one occasion:*

0   10   20   30   40   50   60   70   80   90   100

b) *Have you ever engaged in this practice?*

- Never   □
- Once   □
- Occasionally   □
- Frequently   □
- Almost always   □

c) *What's your option of this practice?*

- It should be used almost always   □
- It should be used often   □
- It should only be used rarel   □
- It should never be used   □

d) *Optional Question: Why do you think this practice should or shouldn't be used?*

_____

_____

_____

Deciding to exclude data points after first checking the impact on statistical significance (e.g., $p \leq 0.05$) or some other desired statistical threshold.

a) *Please estimate the percentage of researchers who you believe have engaged in this practice on at least one occasion:*

0   10   20   30   40   50   60   70   80   90   100

b) *Have you ever engaged in this practice?*

- Never   □
- Once   □
- Occasionally   □
- Frequently   □
- Almost always   □

c) *What's your option of this practice?*

- It should be used almost always   □
- It should be used often   □
- It should only be used rarely   □
- It should never be used   □

d) *Optional Question: Why do you think this practice should or shouldn't be used?*

_____

_____

_____

Collecting more data for a study after first inspecting whether the results are statistically significant (e.g., $p \leq 0.05$).

a) *Please estimate the percentage of researchers who you believe have engaged in this practice on at least one occasion:*

0   10   20   30   40   50   60   70   80   90   100

b) *Have you ever engaged in this practice?*

- Never   □
- Once   □
- Occasionally   □
- Frequently   □
- Almost always   □

c) *What's your option of this practice?*

- It should be used almost always   □
- It should be used often   □
- It should only be used rarely   □
- It should never be used   □

d) *Optional Question: Why do you think this practice should or shouldn't be used?*

_____

_____

_____

Changing to another type of statistical analysis after the analysis initially chosen failed to reach statistical significance (e.g., $p \leq 0.05$) or some other desired statistical threshold.

a) *Please estimate the percentage of researchers who you believe have engaged in this practice on at least one occasion:*

0   10   20   30   40   50   60   70   80   90   100

b) *Have you ever engaged in this practice?*

- Never   □
- Once   □
- Occasionally   □
- Frequently   □
- Almost always   □

c) *What's your option of this practice?*

- It should be used almost always   □
- It should be used often   □
- It should only be used rarely   □
- It should never be used   □

d) *Optional Question: Why do you think this practice should or shouldn't be used?*

_____

_____

_____

Not disclosing known problems in the method and analysis, or problems with the data quality, that potentially impact conclusions.

a) *Please estimate the percentage of researchers who you believe have engaged in this practice on at least one occasion:*

0   10   20   30   40   50   60   70   80   90   100

b) *Have you ever engaged in this practice?*

- Never   □
- Once   □
- Occasionally   □
- Frequently   □
- Almost always   □

c) *What's your option of this practice?*

- It should be used almost always   □
- It should be used often   □
- It should only be used rarely   □
- It should never be used   □

d) *Optional Question: Why do you think this practice should or shouldn't be used?*

_____

_____

_____

Filling in missing data points without identifying those data as simulated.

a) *Please estimate the percentage of researchers who you believe have engaged in this practice on at least one occasion:*

0   10   20   30   40   50   60   70   80   90   100

b) *Have you ever engaged in this practice?*

- Never   □
- Once   □
- Occasionally   □
- Frequently   □
- Almost always   □

c) *What's your option of this practice?*

- It should be used almost always   □
- It should be used often   □
- It should only be used rarely   □
- It should never be used   □

d) *Optional Question: Why do you think this practice should or shouldn't be used?*

_____

_____

_____

Researcher Integrity

Have you ever had doubts about the scientific integrity of research?

	Mild issues/Questionable Research Practices			Serious issues/Scientific Misconduct		
	Never	*Once or twice*	*Often*	*Never*	*Once or twice*	*Often*
Research from other institutions	□	□	□	□	□	□
Research at your institution	□	□	□	□	□	□
Graduate student research at your institution	□	□	□	□	□	□
Senior colleagues and/or collaborators	□	□	□	□	□	□
Your own research	□	□	□	□	□	□

Demographic Inf

Are you a...

- Graduate student   □
- Postdoctoral fellow   □
- Mid-career research fellow/academic   □
- Senior research fellow/academic   □

What is your gender?

- Male   □
- Female   □
- Non-binary/ third gender   □

What is your age?

- Under 20   □
- 20–29   □
- 30–39   □
- 40–49   □
- 50–59   □
- 60–69   □
- 70+   □

What is your sub-discipline? _____
Thank you for participating. Before you leave, do you have any comments, questions, or suggestions about this survey? If so, please enter them below

_____
_____
_____

Source: Fraser et al. (2018).

7 Addressing the reproducibility crisis

Abstract

For a study's results to be confirmed as correct identical results need to occur after the same analytical procedures have been applied to the same dataset. Despite the value of authenticating the truthfulness of a study's results, it is standard practice not to preserve or share datasets or to adequately describe the analytical processes used so that a repeat analysis can occur. These practices have contributed to the development of a 'reproducibility crisis'. This chapter begins with a description of reproducibility and a summary of irreproducible research within different disciplines. The potential economic and social consequences of irreproducible research are then outlined. To mitigate the impact of these consequences, strategies that can increase reproducible research are then suggested, including addressing the barriers to publishing datasets and establishing journals that publish reproducible studies. The purpose of this chapter is to improve the likelihood of reproducible research being produced, which will safeguard the integrity of results that will be produced.

Keywords: Datasets; Novel Results; Open Science Badges; Reproducibility Crisis

Key points

- The reproducibility crisis is caused by the inability to verify that the results of a study are truthful. This crisis is caused by researchers destroying datasets once their study has been published and/or publishing inadequate descriptions about how the dataset was analysed.
- The reproducibility crisis can undermine the public's confidence and investment in research.
- Several strategies should be used to mitigate the reproducibility crisis, including journals requiring authors to submit datasets along with their manuscripts for peer review, journals using open science badges, and establishing journals that only publish studies that are reproductions of previous studies.

7.1   Defining reproducibility

The American president Ronald Reagan, in negotiation with his Soviet counterpart Mikhail Gorbachev, frequently invoked a Russian proverb that means "trust, but verify" ('Doveryai, no proveryai'). At one time, perhaps, there was little reason to suspect that the sciences could not be trusted and that verification was necessary. The replication crisis, along with the relatively new phenomenon of scientists being spectacularly wrong in

DOI: 10.4324/9781003510376-7

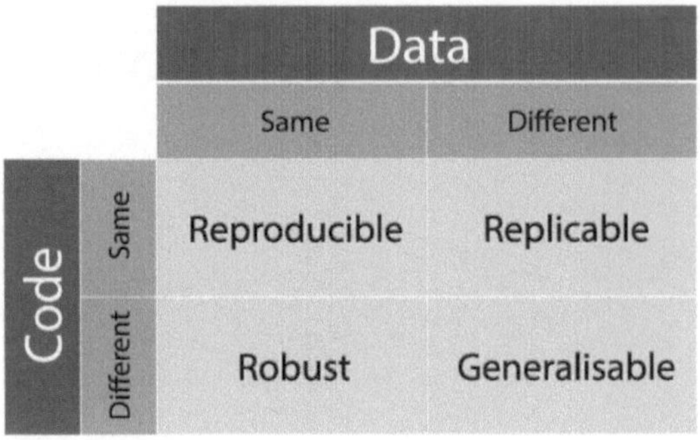

Figure 7.1 Reproducible, robust, replicable, and generalisable research
Source: Leipzig et al. (2021, p. 2).

> public (as we saw with some high-profile errors in COVID-19 research), has highlighted the crucial role of verification in buttressing the credibility of science.
>
> (Vazire & Holcombe, 2021, p. 9)

As explained above, historically the public have unquestionably considered knowledge and scientific discoveries as truthful. However, since the late 2010s, studies have cast doubt on that assumption. This is because in many studies, published results cannot be verified and confirmed as truthful. The main cause of this lack of trust in science is that the study's dataset is unavailable and/or the study's methodology was inadequately described meaning that a repeat analysis of the dataset could not be competently performed (see Table 7.1). The inability to verify a study's results has been termed the 'reproducibility crisis'. However, there are some who believe that this inability is not a crisis (Munafò et al., 2022). Since 2015 there has been a rapid increase in the number of studies about the reproducibility crisis (Wass et al., 2019).

Understanding the distinctions between 'reproducibility', 'robustness', 'generalisability', and 'replicability' can help contextualise the reproducibility crisis. *Reproducible* refers to confirming if a study's results are correct by repeating an analysis of the study's dataset using the same analytical tools and procedures (i.e., code). *Replicable* means using the same analytical tools and procedures to examine different datasets. *Robust* is defined as using the same dataset but different analytical tools and procedures. Finally, *generalisable* refers to analysing different datasets with different analytical tools and procedures (Hejblum et al., 2020; Leipzig et al., 2021) (see Figure 7.1). The proportion of irreproducible research has been measured in other disciplines, such as addiction medicine and neurology research (see Table 7.1).

Serghiou et al. (2021) conducted one of the most comprehensive examinations about transparency and reproducible factors within biomedicine. They examined the entire open access biomedical literature on PubMed central (i.e., 2.75 million articles) using an automated search program that identified five indicators of transparency and reproducibility, which were data sharing, code sharing, conflicts of interest disclosures, funding disclosures, and protocol registration. They discovered that conflicts of interest and funding disclosures have increased while protocol registration and code sharing have not (see Figure 7.2).

7.2   Consequences of irreproducible research

For some their confidence in the integrity of research can be undermined by irreproducible research. Mede and colleagues (2021) reported that most German people that they surveyed did not consider irreproducible research to be detrimental to scientific endeavours since efforts to guarantee that results could be reproduced were considered a part of science's self-correcting nature. However, those who supported AfD, the populist right-wing political party, considered failures to replicate a study's results as evidence for their already established 'anti-scientific sentiments' (Mede et al., 2021).

Table 7.1 Proportion of irreproducible research by discipline

Reference	*Discipline*	*Results*	*Conclusion (Excerpt)*
Adewumi et al. (2021)	Addiction medicine	Analysis script availability: 2/244, 0.82% Conflict of interest statement: 221/293, 75.43% Data availability: 28/244, 11.48% Material availability: 2/237, 0.84% Open access publications: 152/300, 50.70% Pre-registration: 7/244, 2.87% Protocol availability: 3/244, 1.23% Replication study identified: 1/244, 0.4% Statement about funding sources: 268/293, 91.47%	'Our study found that current practices that promote transparency and reproducibility are lacking, thus, there is room for improvement. In particular, investigators should pre-register studies prior to commencement. Researchers should also make the materials, data, and analysis script publicly available. Further, individuals should be transparent about funding sources for the project and financial conflicts of interest' (Adewumi et al., 2021, p.1).
Fladie et al. (2019)	Nephrology literature	Analysis script availability: 4/172, 2.3% Data availability statements: 43/172, 25% Open access publications: 152/300, 50.7% Pre-registration: 8/71, 11.3% Protocol availability: 0/71, 0.0%	'"Our study found that reproducible and transparent research practices are infrequently used by the nephrology research community. Greater efforts should be made by both funders and journals. In doing so, an open science culture may eventually become the norm rather than the exception.' (Fladie et al., 2019, p. 173).
Johnson et al. (2020)	Otolaryngology	Analysis script availability: 0.6% Data availability: 2% Pre-registration: 3.9% Protocol availability: 5.3%	'Inadequate reproducibility practices exist in otolaryngology. Nearly all studies in our analysis lacked a data or material availability statement, did not link to an accessible protocol, and were not preregistered. Taking steps to improve reproducibility would likely improve patient care.' (Johnson et al., 2020, p. 1894).
Okonya et al. (2020)	Anaesthesiology research	Analysis script availability: 1/188, 0.53% Statement about data availability: 24/188, 12.76% Material availability: 28/188, 14.89%	'Anaesthesiology research needs to improve indicators of reproducibility and transparency. By making research publicly available and improving accessibility to detailed study components, primary research can be reproduced in subsequent studies and help contribute to the development of new practice guidelines.' (Okonya et al., 2020, p. 835).

(Continued)

Table 7.1 (Continued)

Reference	Discipline	Results	Conclusion (Excerpt)
Rauh et al. (2020)	Neurology research	Analysis script availability: 0.7% Data availability: 9.2% Material availability: 9.4% Open access publications: 97.25% Pre-registration: 3.7% Protocol availability: 0.7%	'Currently, published neurology research does not consistently provide information needed for reproducibility. The implications of poor research reporting can both affect patient care and increase research waste. Collaborative intervention by authors, peer reviewers, journals, and funding sources is needed to mitigate this problem.' (Rauh et al., 2020, p. 1).
Rauh et al. (2022)	Urology research	Protocol availability: 0.58% Access to raw data: 4.09% Access to materials: 3.09% Pre-registered studies: 4.68% Availability of analysis scripts: 0%	'Current urology research does not consistently provide the components needed to reproduce original studies. Collaborative efforts from investigators and journal editors are needed to improve research quality while minimising waste and patient risk.' (Rauh et al., 2022, p. 1).
Wallach et al. (2018)	Biomedical literature	Conflict of interest statement: 97/149, 65.1% Data availability: 19/104, 18.3% Protocol availability: 1/104, 0.96% Replication study identified: 5/97, 5.2% Statement about funding sources: 103/149, 69.1%	'Our evaluation suggests that although there have been improvements over the last few years in certain key indicators of reproducibility and transparency, opportunities exist to improve reproducible research practices across the biomedical literature and to make features related to reproducibility more readily visible in PubMed.' (Wallach et al., 2018, p. 1).
Walters et al. (2019)	Oncology	Conflict of interest statement: 65/194, 25% Data availability: 9/194, 4.6% Pre-registration statements: 7/194, 3.60% Protocol availability: 5/194, 2.6%	'We found that key reproducibility and transparency characteristics were absent from a random sample of published oncology publications. We recommend required preregistration for all eligible trials and systematic reviews, published protocols for all manuscripts, and deposition of raw data and metadata in public repositories.' (Walters et al., 2019, p. 1).
Sherry et al. (2020)	Psychiatry	Analysis script availability: 1/296, 0.33% Conflict of interest statement: 177/296, 59.79% Data availability: 1/296, 0.33% Material availability: 17/296, 5.74% Open access publications: 107/296, 36.14% Protocol availability: 4/296, 1.35% Statement about funding sources: 185/296, 62.5%	'Currently, Psychiatry research has significant potential to improve adherence to reproducibility and transparency practices. Thus, this study presents a reference point for the state of reproducibility and transparency in Psychiatry literature. Future assessments are recommended to evaluate and encourage progress.' (Sherry et al., 2020, p. 1).

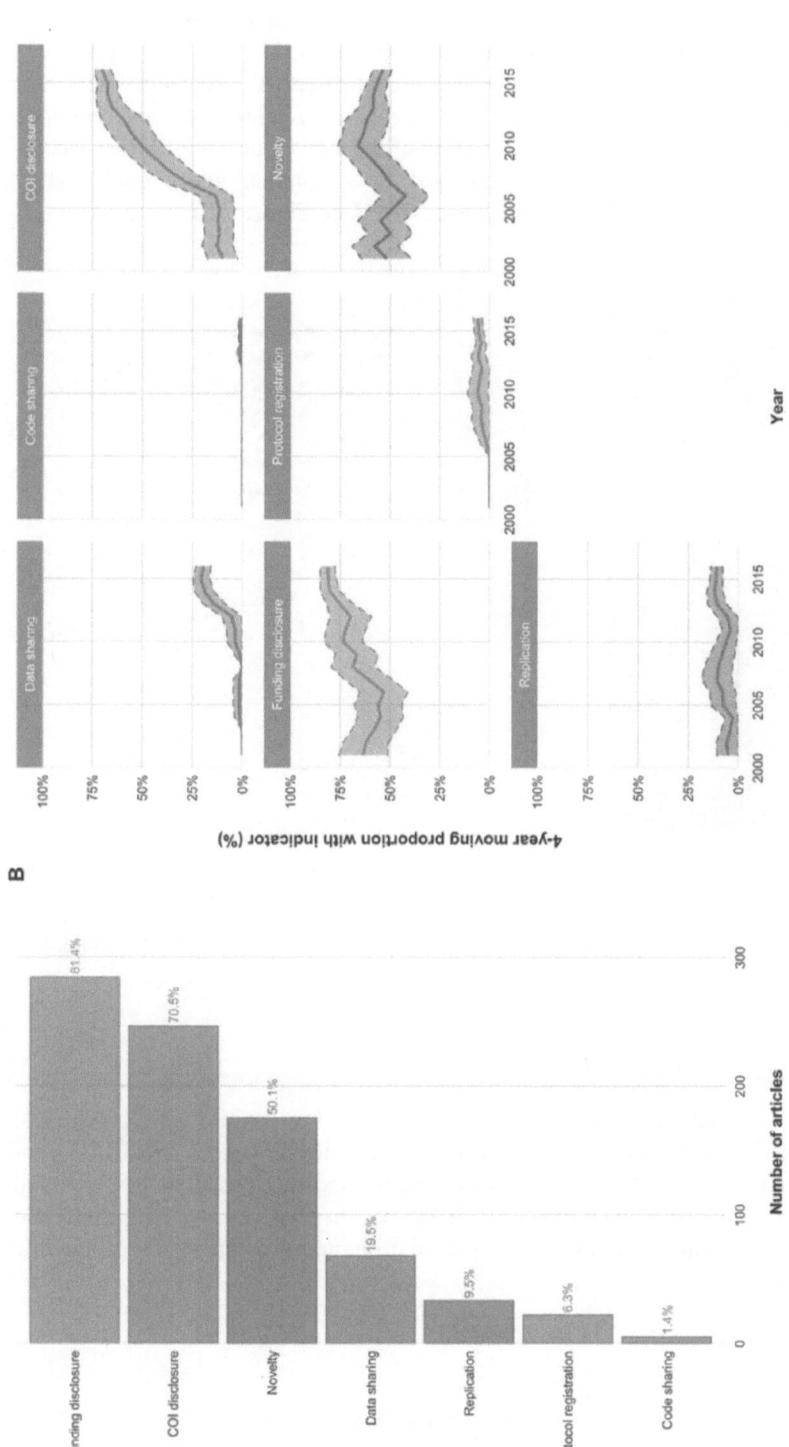

Figure 7.2 Indicators of transparency and reproducibility within 349 biomedical publications from 2000 to 2015

Source: Serghiou et al. (2021, p. 4).

Note: (A) Indicators of transparency and reproducibility across 349 research articles (2015–2018). Most publications included Conflict of Interest or Funding disclosures, but few mentioned Data, Code, or Protocol sharing. Similarly, most claimed Novelty but few mentioned a Replication component.

(B) Indicators of transparency and reproducibility across time on the basis of manual assessment. These graphs merge data from this study on 349 research articles (2015–2018) with similar data from two previous studies on another 590 PubMed articles (2000–2018). Proportions are displayed as a four-year centred moving average. The shaded region indicates the 95% Confidence Interval. The most notable change is that of Conflict of Interest disclosures, the reporting of which increased from 12% in 2000 to 76% in 2018. The data underlying this figure can be found on Open Science Framework at http://www.doi.org/10.17605/OSF.IO/E58WS.

Irreproducible research can also result in a wastage of financial investment that has been allocated to research. Freedman et al. (2015) estimated that annually USD 28.2 billion was spent in the United States of America on irreproducible preclinical research. The main factor that caused research to be irreproducible was the inability to access biological reagents and reference materials (36.1%), followed by an inadequate description of the study's design (27.6%), lack of a detailed description about how the dataset was analysed (25.5%), and a failure to adequately describe laboratory protocols (10.8%) (Freedman et al., 2015) (see Figure 7.3).

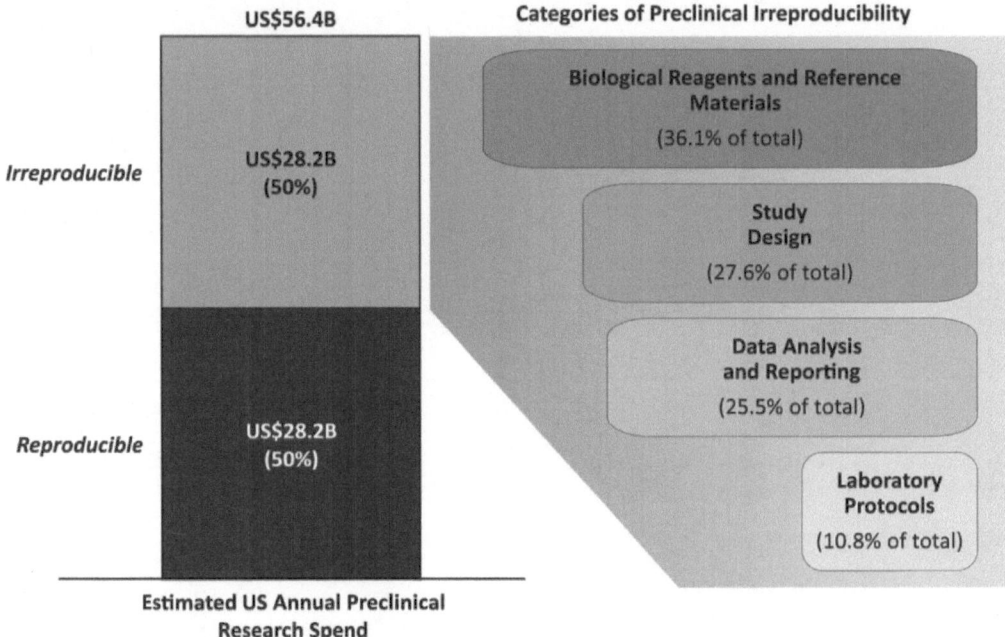

Figure 7.3 Estimated cost of preclinical research and categories of errors that contribute to irreproducibility in the United States of America

Source: Freedman et al. (2015, p. 4).

7.3　Strategies to increase reproducible research

There are two overarching recommendations that can improve the creation of reproducible research. First, all datasets, tools, procedures, protocols, and analysis scripts should be published and clearly explained. Second, all sources of funding and any potential or perceived conflicts of interest should be declared (Rauh et al., 2020; Wright et al., 2020). With these two overarching recommendations in mind, seven strategies to promote the creation of reproducible research are explained below.

7.3.1　*Publishing datasets*

7.3.1.1　*Conventional practices about publishing and archiving datasets*

To confirm that a study's results are correct identical results need to be created after the same dataset has been reanalysed using the original study's analytical tools and procedures. Failure to preserve a study's dataset for this purpose can result in the results always being doubted.

Currently, it is conventional practice to destroy datasets once a study has been published. Ioannidis (2012, p. 646) once wrote about this practice:

> … it is also possible that a Library of Alexandria actually disappears every few minutes. Currently, there are petabytes of scientific information produced on a daily basis and millions of papers are being published annually. In most scientific fields, the vast majority of the collected data, protocols, and analyses are not available and/or disappear soon after or even before publication. If one tries to identify the raw data and protocols of papers published only 20 years ago, it is likely that very little is currently available. Even for papers published this week, readily available raw data, protocols, and analysis codes would be the exception rather than the rule.

Ioannidis's sentiment has also been confirmed by Minocher et al. (2021), who attempted to obtain datasets for 560 empirical studies that were published between 1955 and 2018 that examined either animal behaviours, behavioural ecology, cultural evolution, or evolutionary psychology. They reported that datasets could only be recovered for 30% of all articles in the sample. They also discovered that the ability to retrieve datasets for studies exponentially declined after the study's publication, halving every nine years for experimental data involving humans (Minocher et al., 2021) (see Figure 7.4).

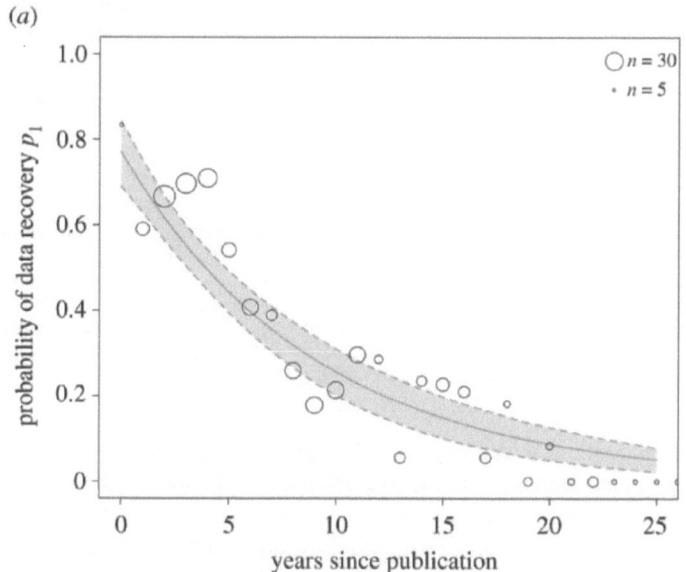

Figure 7.4 Probability of data recovery since a study's publication

Source: Minocher et al. (2021, p. 6).

Note: The predicted probability of data recovery declines exponentially with increasing time since publication, halving every six years. The solid line plots the expected exponential decay curve and the shaded interval between dotted lines shows the 89% compatibility interval. The empty circles plot the observed data, that is, raw proportion of studies, for each year, for which we obtained materials. The size of each circle is scaled by the total number of observations for that year. The x-axis is truncated at 26 years since publication, for clarity of presentation, as the expected probabilities beyond this point are <0.01.

7.3.1.2   Researchers' attitudes about disseminating datasets

To reverse the practice of destroying datasets understanding a researcher's attitudes about preserving their datasets is essential. Borghi and Van Gulick (2021) surveyed 274 psychology researchers

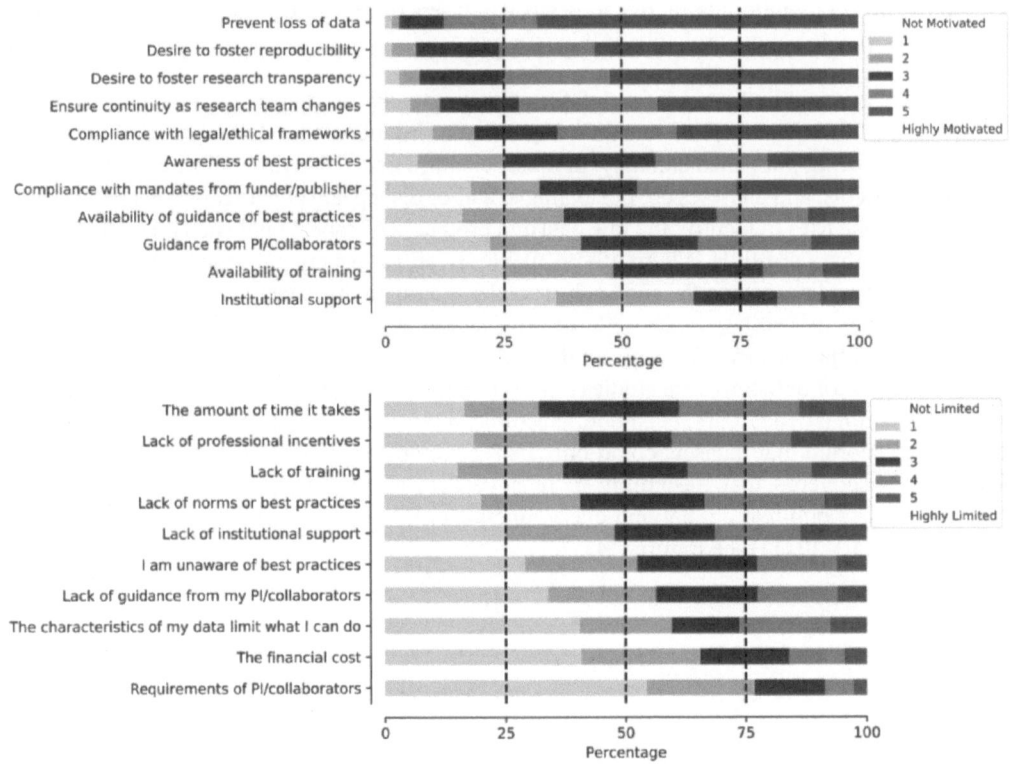

Figure 7.5 Factors that motivate and limit psychology researchers' data management practices

Source: Borghi and Van Gulick (2021, p. 6).

Note: Participants were asked to rate the degree to which different factors motivate and limit their data-related practices on a scale of 1 (not limited/motivated) to 5 (highly limited/motivated). For motivations, participants gave high ratings to immediate and practical concerns, such as the desire not to lose data as well as broader concerns such as the desire to foster reproducibility and research transparency. Ratings for limitations were more diffuse.

from 31 countries about their data management attitudes. They reported that most expressed motivation to implement data management practices to prevent the loss of data and to foster reproducibility and transparent research practices. However, many reported that they did not engage in data management practices, citing financial costs and the requirements of collaborators (see Figure 7.5).

Mozersky et al. (2021) asked 425 respondents in the United States of America to describe both the barriers and enablers that influenced their decisions to share their study's qualitative data. The most common concern expressed was a 'lack of permission from research participants to share data' ($n=370$, 86%). In contrast, the least common concern was 'I do not like the idea of others judging my work' ($n=74$, 17%) (see Table 7.2).

Mozersky et al. also asked respondents what factors would increase their willingness to share their data. The most common factor discovered was to 'increase the societal impact of research' ($n=353$, 83%) while the least common factor was 'if journals required data to be shared' ($n=201$, 47%) (see Table 7.3).

Mozersky et al. also reported what resources would increase the respondent's willingness to share their data. Most believed that they would share such data if funding agencies would pay for the costs associated with establishing a data sharing repository ($n=294$, 69%), if they were

Table 7.2 Concerns regarding qualitative data sharing

How concerned are you about the following factors related to the idea of sharing qualitative data through a repository?	Frequency of participants who indicated concern (%)
Lack of permission from research participants to share data	370 (87%)
The sensitivity of research data	360 (85%)
Breach of trust with participants	349 (82%)
IRB or institutional policies	336 (79%)
Concern that data cannot be adequately anonymised	334 (79%)
Losing control over who has access to my qualitative data	326 (77%)
The time and effort to prepare data for deposit	325 (76%)
The potential for misinterpretation of my data by other researchers	316 (74%)
Financial cost to prepare qualitative data for deposit	283 (67%)
Issues with legal permissions	252 (59%)
Potential for repository technology failure	233 (55%)
My lack of knowledge about repositories and data sharing in general	223 (52%)
Others do not deserve to use data I collected	94 (22%)
I do not like the idea of others judging my work	74 (17%)

Source: Mozersky et al. (2021, p. 7).

Note: Items were rated on a scale of 1 (not at all concerned), 2 (slightly concerned), 3 (moderately concerned), 4 (very concerned), or 5 (extremely concerned). Participants were considered to be concerned if they rated 3, 4, or 5.

Table 7.3 Considerations that would increase willingness to share data

How likely would each of the following considerations increase your willingness to share qualitative data through a repository?	Frequency of participants willing to share (%)
If sharing increased the societal impact of research	353 (83%)
If I knew my participants would agree to data sharing	339 (80%)
If sharing led to increased collaborations	322 (76%)
If sharing decreased the burden on participant communities	308 (72%)
If secondary data users needed to cite their data sources in all publications	294 (69%)
If data could be reused to explore new research questions	283 (67%)
If sharing made data from publicly-funded research more widely available	286 (67%)
If repositories provided a secure infrastructure for data storage	279 (66%)
If those who share data are invited to be co-authors on papers that use data	275 (65%)
If funding agencies required data to be shared	266 (63%)
If sharing helped avoid duplication of work	260 (61%)
If sharing data created the opportunity for students to learn how to analyse data	257 (60%)
If sharing allowed for verification of data interpretation	230 (54%)
If sharing positively influenced career promotion decisions	226 (53%)
If repositories provided a central catalogue of available data sets	214 (50%)
If sharing led to increased citations	205 (48%)
If journals required data to be shared	201 (47%)

Source: Mozersky et al. (2021, p. 7).

Note: The degree to which each consideration would increase willingness to share qualitative data were rated on a scale of 1 (not at all likely), 2 (somewhat unlikely), 3 (neutral), 4 (somewhat likely), or 5 (very likely). Participants were considered willing to share if they rated 4 or

given clear guidance on ethics and compliance-related issues (*n*=259, 61%), and if the data deposited was anonymised (*n*=243, 57%). In contrast, less than half believed that they would be willing to share their data if the data repository provided consultations regarding the sharing of data (*n*=207, 49%).

7.3.1.3   Teaching academics data management processes

Researchers should develop the skills required to use effective data management to ensure that their datasets are archived. Appropriate data archival practices will give others the opportunity to conduct reproducible studies and/or combine the archived dataset with their own dataset to improve the generalisability of their findings. Kanza and Knight (2022) have outlined ten strategies that can help researchers perfect their data management practices (see Table 7.4).

Table 7.4 Ten strategies for good research data management

Tip	Description
Start early	Plan your data management strategy right from the start, think about every aspect of your project from the data collection, organisation, storage, and even where you are planning on publishing and sharing the results.
Data management plan	Make one of these right at the beginning and refer to it and improve it throughout the entire project life cycle.
Organisation is key	Use sensible folder/file structures that have been agreed with the entire team.
Version control your work	Decide on what version control systems you are going to use and implement these plans from the beginning.
Storage strategy	Consider your long term and short term data storage. And implement the 321 data storage rule: (three copies of the data, within two types of media, with one stored at a separate site), and NEVER rely on USB sticks.
Remember your standards and be FAIR	Think about what standards you are going to make your data available in. Data should be Findable, Accessible, Interoperable and Re-useable (FAIR).
Consider ethics	If you are interacting with human data in any way, you will need ethics! These applications can take a while to write and obtain approval for, so start straight away!
Factor in resources	Time and costs should be factored in for all required resources, including your data management!
Future proof your data	Metadata alone will not future proof your data, you should get DOI's for your datasets and include relevant README's and description files.
Communicate	If you are working on collaborative research projects then communication is key both in setting up the initial organisational strategies, and throughout the entire project life cycle to ensure that team members are working consistently with respect to data collection, organisation, storage etc.

Source: Kanza and Knight (2022).

7.3.1.4   Improving international data sharing agreements

Sometimes modifying international data sharing agreements can jeopardise the act of obtaining a dataset to repeat an analysis to confirm existing results. Such agreements sometimes change after one of the partners changes their policies about data management and/or data sharing. Additionally, within a jurisdiction legislation concerning data sharing can change, preventing it from occurring. This dilemma was once described by Devriendt et al. (2021, p. 7), who stated:

> We have had collaboration with another research group in Australia. Recently, the data access agreement expired (…) and we tried to renew it. The legal team of [our institute] said that it needs to be the controller-to-controller version of the EU contract clauses and the legal team of the other institute said that they could not sign such an agreement. There were

disagreements basically on the [precise wording] of the agreement, but [our institute] cannot easily change the wording of these clauses. We had to cut the data access for the Australian team although we have worked together without problems for ten years.

Two strategies can be used to reduce the prospect of international data sharing agreements being cancelled. First, such agreements should contain grandfather clauses that ensure that any data already collected is not subjected to the changing terms of the agreement. Second, the possibility of the dataset and agreement cancelling during the study should be debated and discussed before the study begins. Such deliberations could yield viable solutions in the event of data sharing agreements being cancelled.

7.3.1.5   *Journal requesting datasets from authors*

Traditionally, when an author submits a manuscript for peer review, they are not required to also submit the corresponding dataset. Furthermore, some aspiring or accomplished authors are unwilling to provide their datasets when a journal's editor requests this information. Recalling his role as the editor-in-chief of *Molecular Brain*, from early 2017 to September 2019, Miyakawa (2020) reviewed 180 manuscripts for publication. During his appointment he requested datasets for 41 submitted manuscripts. Of these, 21 were withdrawn without providing any raw data and the other 20 were resubmitted with some raw data. Of these 20 resubmissions, 19 were rejected due to insufficient raw data and/or there was a mismatch between the raw data and the results. Of the 21 submissions that were withdrawn without providing raw data, 14 were published in other journals. Of these, 12 were published in journals that either required or recommended raw data be provided upon request. Miyakawa mostly received no response ($n=10$) after sending a request for raw data to the authors of those 12 manuscripts (see Figure 7.6). Based on these findings, Miyakawa (2020, p. 1) concluded that:

> Considering that any scientific study should be based on raw data, and that data storage space should no longer be a challenge, journals, in principle, should try to have their authors publicize raw data in a public database or journal site upon the publication of the paper to increase reproducibility of the published results and to increase public trust in science.

The reluctance of authors to provide their study's dataset to journals has also occurred in the journal *BMJ* (i.e., formerly the *British Medical Journal*). As illustrated in Figure 7.7, despite mandating that all datasets must be submitted along with the manuscript, most submissions to *BMJ* have not adhered to this rule (see Figure 7.7).

Aside from *BMJ* and *Molecular Brain*, the reluctance of authors to share their datasets also occurs in other journals. Tedersoo et al. (2021) measured the proportion and reasons why authors who had published manuscripts in nine disciplines in *Nature* and *Science* magazines declined to share their datasets. They reported that the lack of time to search for data (29.2%), loss of data (27.7%), and privacy or legal concerns (23.1%) were the most cited reasons why authors did give these journals their datasets. They also reported that authors who published manuscripts in journals about ecology and forestry ignored requests for data more often than authors who published manuscripts in journals about materials for energy and catalysis, microbiology, psychology, and social sciences (see Figure 7.8) (Tedersoo et al., 2021).

Providing a dataset with a submitted manuscript should be a common practice so that the results in the manuscript can be validated. Academic publishers should mandate that editor-in-chiefs should only receive for peer review manuscripts that also have corresponding

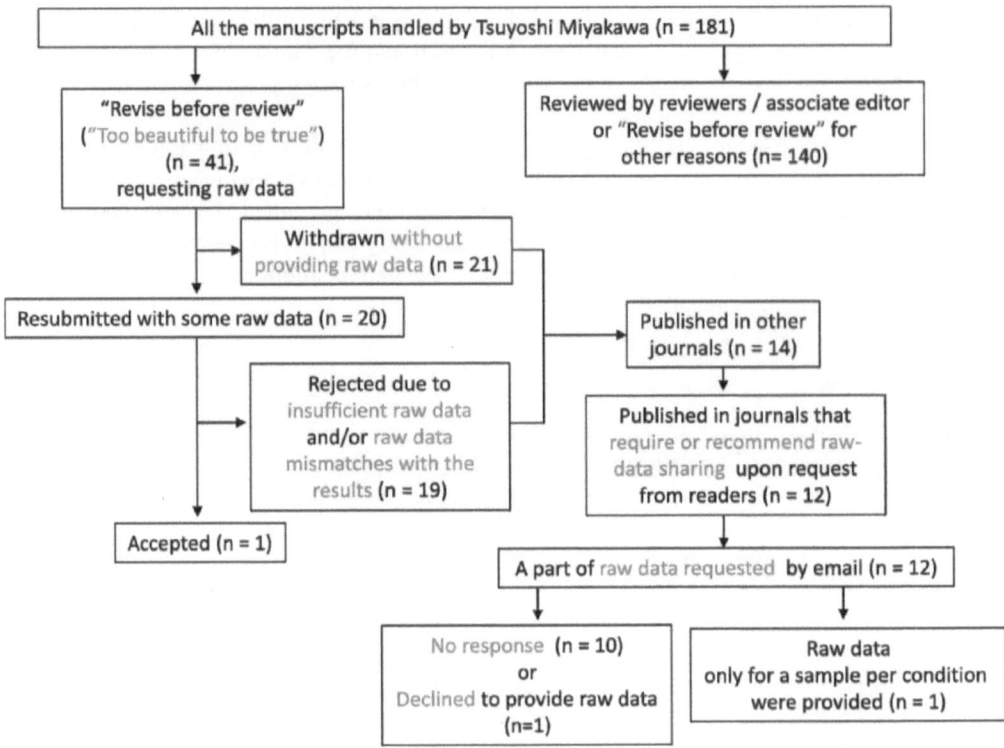

Figure 7.6 Flowchart of the manuscripts handled by Miyakawa in Molecular Brain from December 2017 to September 2019

Source: Miyakawa (2020, p. 2).

datasets. Failure to enforce such a policy will result in the continual publication and proliferation of manuscripts whose results cannot be validated.

7.3.2   *Establishing journals that only publish replication studies*

During 2018, Wallach et al. compared the reproducibility, transparency, and open access factors in biomedical studies that were published between 2000 to 2014 ($n=441$) and 2015 to 2017 ($n=149$). To achieve this task, they assessed each study for the following five factors:

1 Based on the abstract and/or introduction, the index article claims that it presents some novel findings.
2 Based on the abstract and/or introduction, the index article clearly claims that it is a replication effort trying to validate previous knowledge, or it is inferred that the index article is a replication trying to validate previous knowledge.
3 Based on the abstract and/or introduction, the index article claims to be both novel and to replicate previous findings.
4 No statement or an unclear statement in the abstract and/or introduction about whether the index article presents a novel finding or replication.
5 No distinct abstract and introduction exists (Wallach et al., 2018, p. 17).

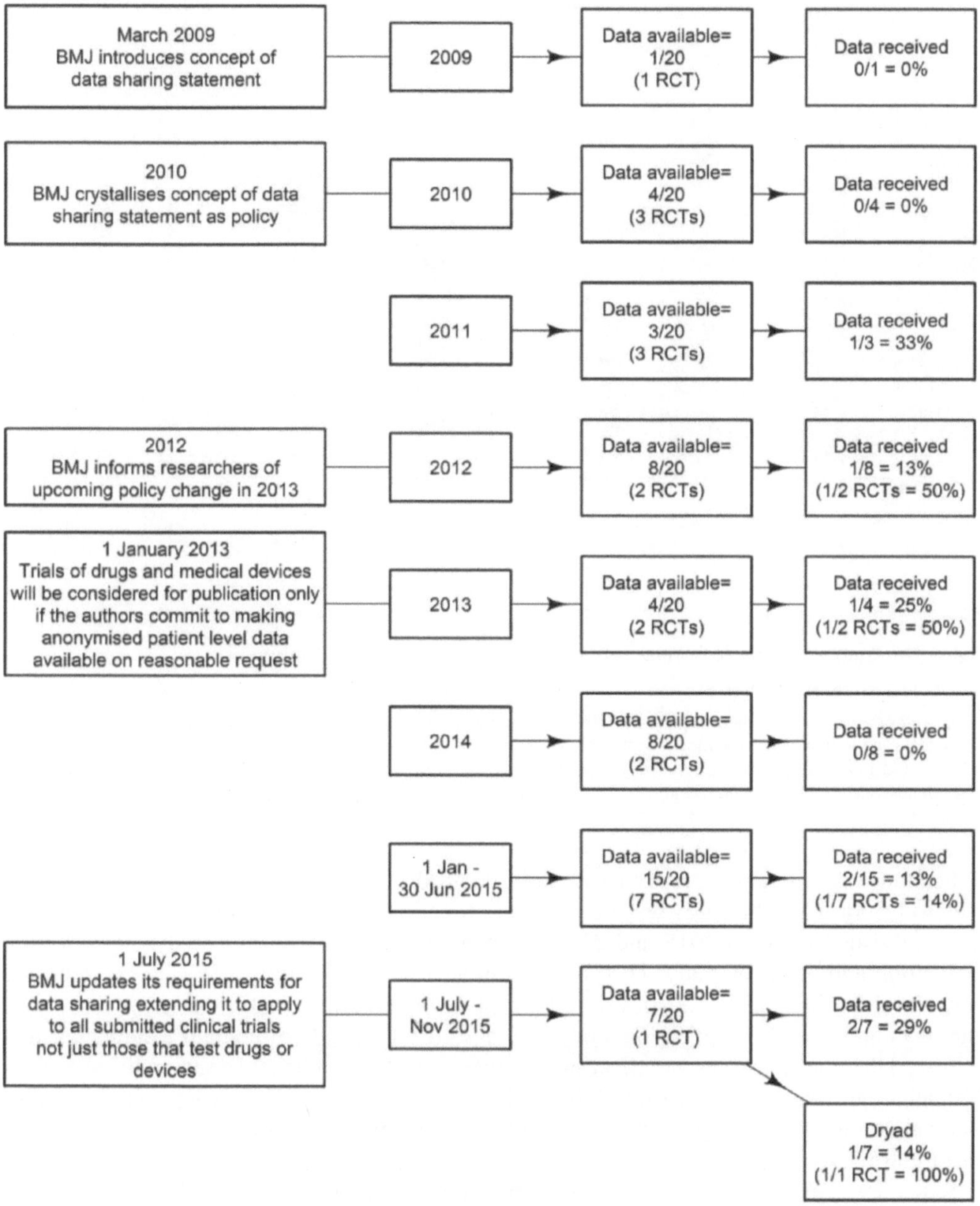

March 2009 BMJ introduces concept of data sharing statement	2009	Data available= 1/20 (1 RCT)	Data received 0/1 = 0%
2010 BMJ crystallises concept of data sharing statement as policy	2010	Data available= 4/20 (3 RCTs)	Data received 0/4 = 0%
	2011	Data available= 3/20 (3 RCTs)	Data received 1/3 = 33%
2012 BMJ informs researchers of upcoming policy change in 2013	2012	Data available= 8/20 (2 RCTs)	Data received 1/8 = 13% (1/2 RCTs = 50%)
1 January 2013 Trials of drugs and medical devices will be considered for publication only if the authors commit to making anonymised patient level data available on reasonable request	2013	Data available= 4/20 (2 RCTs)	Data received 1/4 = 25% (1/2 RCTs = 50%)
	2014	Data available= 8/20 (2 RCTs)	Data received 0/8 = 0%
	1 Jan - 30 Jun 2015	Data available= 15/20 (7 RCTs)	Data received 2/15 = 13% (1/7 RCTs = 14%)
1 July 2015 BMJ updates its requirements for data sharing extending it to apply to all submitted clinical trials not just those that test drugs or devices	1 July - Nov 2015	Data available= 7/20 (1 RCT)	Data received 2/7 = 29%
			Dryad 1/7 = 14% (1/1 RCT = 100%)

Figure 7.7 Summary of data availability and actual data received for BMJ research articles grouped by year and in relation to data sharing policy changes

Source: Rowhani-Farid and Barnett (2016, p. 6).

Acronyms: BMJ, British Medical Journal; RCT, randomised clinical trial.

Of the 441 biomedical studies that were published between 2000 and 2014, 229 were able to be assessed against the five predetermined factors. Of these, 133 (51.4%) reported novel findings, five (1.9%) reported that the study was a replication, five (1.9%) reported both novel findings and were also replication studies, 111 (42.9%) contained no statement about

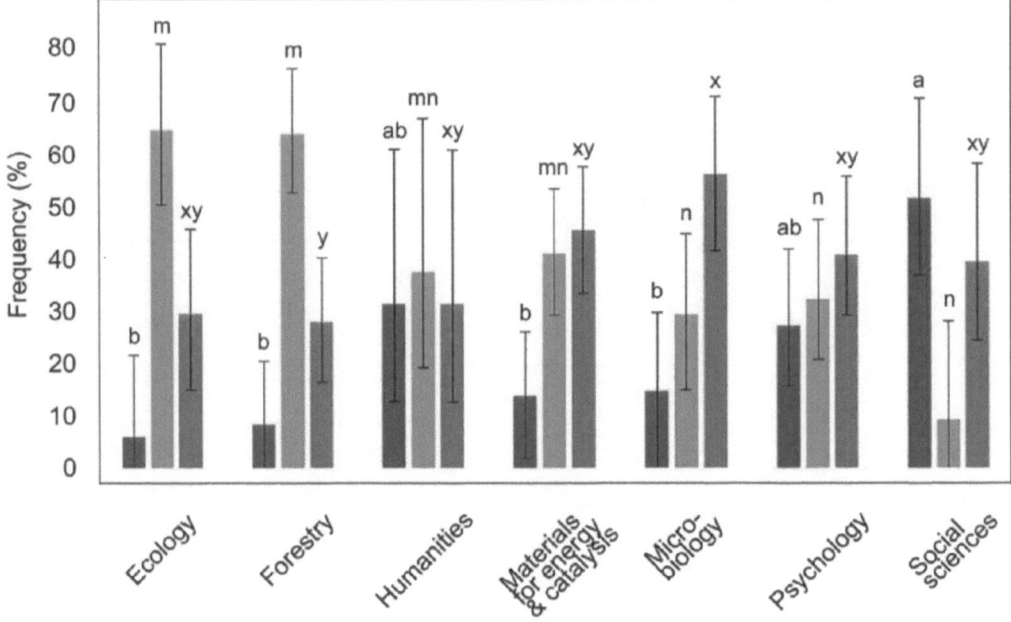

Figure 7.8 Authors' response to data request (*n*=199) depending on discipline

Source: Tedersoo et al. (2021, p. 5).

Note: Authors' response to data request (*n*=199) depending on discipline (blue, declined; orange, ignored; purple, obtained). Bars indicate 95% Confidence interval as determined by Sison and Glaz's statistical test. Letters above bars indicate statistically significant difference groups in frequency of data availability by each category based on Tukey post-hoc test and Bonferroni correction.

if the study was novel or a replication, and five (1.9%) contained no abstract that could be used to determine if the study was novel or a replication. Of the 149 biomedical studies that were published between 2015 and 2017, 97 contained items that indicated if they were a replication and/or novel study. They reported that of these 97 studies, 56 (57.7%) reported only novel findings, five (5.2%) were only replication studies, 10 (10.3%) contained novel findings and were also replication studies, and 26 (26.8%) did not contain a statement that indicated that the study was a replication or contained novel findings. Based on these results, Wallach et al. illustrated that from 2000 to 2017, there has been a tendency to publish bio-medical studies that contained novel findings instead of replications of previous research (see Figure 7.9).

Since few replication studies have been published in biomedical research, and perhaps in other disciplines as well, there is a considerable number of studies that contain novel results that have not been validated using replication. To reverse this trend specialised journals that only publish replication studies should be established within different academic disciplines. Also, to mitigate any actual or perceived conflicts of interest these journals should only accept replication studies written by authors who were not associated with the original study. Publishing replication studies will enable research findings to either be validated or repudiated, which will improve confidence in the validity of knowledge being created and shared.

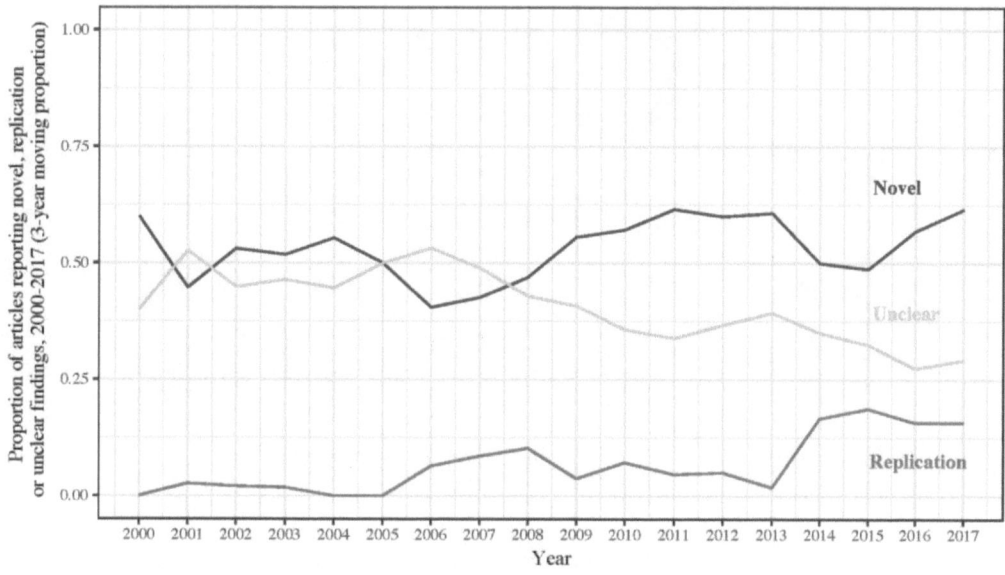

Figure 7.9 Proportion of articles reporting novel, replication, or unclear findings, 2000–2017 (three-year moving proportion)

Source: Wallach et al. (2018, p. 11).

Note: Underlying data for Fig 3 can be found at https://osf.io/3ypdn/.

7.3.3   *Teaching academic staff about reproducibility*

Publishing studies is one of several tasks that academics perform. For some academics this requirement can be onerous because they draft manuscripts outside of conventional work hours, including on weekends (Barnett et al., 2019). Since academics are expected to publish studies, they are not necessarily concerned or motivated with ensuring that the research that they publish is reproducible. To illustrate this concept in his 2012 article, titled 'Why Science Is Not Necessarily Self-Correcting', Ioannidis describes how, on the fictitious planet F345, practices that ensure that studies are reproducible are always subordinate to discovering new findings (see Appendix 7.1). To ensure that academics create studies that are reproducible universities should give academic staff, regardless of their seniority or work experience, training courses that can enhance their skills at creating reproducible research. This sentiment was expressed by Diaba-Nuhoho and Amponsah-Offeh (2021, p. 3), who stated:

> We all must be on the lookout and guard against any act that may undermine integrity of any research work. Going forward, postgraduates, postdocs, supervisors, technicians, laboratory managers and all who are involved in research needs to have compulsory and periodic courses on research integrity throughout their career and must be an integral part of their professional development.

7.3.4   *Open Science Badges*

To increase awareness among academics about the importance of creating reproducible research the *Center for Open Science* (2017) has created five Open Science Badges (see Figure 7.10)

Figure 7.10 Logos of Open Science Badges by the Center for Open Science

Source: American Psychiatric Association (APA) Journals (2022).

(see Appendix 7.2). Since their inception, Open Science Badges have been endorsed and used in several journals, including *The International Journal for the Psychology of Religion* (van Elk et al., 2018), *The Journal of Social Psychology* (Grahe, 2014), and *The Canadian Journal of Experimental Psychology* (Pexman, 2017). According to the *Center for Open Science* Open Science Badges:

> are included on publications and signal to the reader that the content of the publication has been made publicly available and certify its accessibility in a persistent location. They acknowledge open science practices are incentives for researchers to share data, materials, or to preregister protocols and have proven to be successful and continue to gain visibility in the scientific community.
>
> (Kretser et al., 2019, p. 334)

Since their inception Open Science Badges have produced mixed outcomes. Kidwell et al. (2016) and Rowhani-Farid et al. (2020) have both examined if the 'Open Data' Open Science Badge has motivated researchers to provide datasets along with their manuscripts to journals. Rowhani-Farid et al. reported that this badge did not motivate researchers who published in *BMJ Open* to share their data. Kidwell et al., however, reported that when badges were earned the data was more likely to be available, correct, useable, and complete. Despite these mixed results the influence of other Open Science Badges in making studies reproducible, such as the 'Preregistered' badge, have not been measured. To determine the effectiveness of these badges in fostering the practice of making research reproducible more research needs to be conducted. For example, researchers can randomly select a sample of studies with datasets that have a pre-registered badge and repeat an identical analysis. This reanalysis can help determine if the pre-registered badge improved the description of the study and, by extension, its reproducibility.

7.3.5   *Incorporating reproducibility requirements into the criteria for research funding*

We believe it is time for a fundamental cultural shift in the scientific community: rigor and reproducibility should become primary concerns in the criteria and decision-making process of designing studies, funding research, and writing and publishing results. Successful systematic adoption of best practices will require the buy-in of multiple stakeholders in scientific communities: publishers, academic institutions, funding agencies, and stakeholders.

(Brito et al., 2020, p. 5)

As explained in the quotation above, multiple stakeholders need to change their practices so that there is a greater chance of research being reproducible. Funding agencies also have a role to play in making research reproducible. However, as explained in Chapter 4, Diong et al. (2021) showed that agencies in Australia that fund medical research rarely require applicants to use mechanisms to ensure that their study is reproducible. To address this issue agencies that fund research should model their funding criteria on international agencies that mandate that researchers produce research that is reproducible (Kowalczyk et al., 2022). Additionally, they should implement two strategies that are crucial for reproducibility. First, before they collect data researchers should be obligated to publicly register their study's design. This act will improve the clarity of the study design and will increase the prospect of the study being reproducible. Second, when the study is finished the researchers should be legally required to share their datasets in a deidentified form upon reasonable request so that a reanalysis of the dataset can confirm the results.

7.3.6   *Reforming academic hiring practices to promote reproducible research*

It is important that institutions ensure that organisational structures within which researchers work reward engagement with and adoption of open and transparent research practices. Academic hiring decisions, annual performance reviews, and promotion are often informed by easy-to-calculate research metrics such as the number of research outputs an academic has produced, or the amount of grant income an academic has generated within a particular period. A high score on these metrics does not mean that the underlying research is transparent and robust (often simply that there is a lot of it). Academics need to be incentivised to produce research that is both high-quality and transparent.

(Stewart et al., 2021, pp. 2–3)

As described in the above quotation, obtaining an academic job, promotion, or tenure is primarily influenced by the number of articles that a candidate has published and the number of times that their article has been cited. Consequently, there has not been much emphasis on acknowledging, let alone rewarding with career opportunities, researchers that incorporate reproducibility practices into their research activities. Academic institutions should include a criterion about reproducible research practices for academic positions. This change of job criteria could contribute to a cultural change within academic institutions, away from rewarding academics who have favourable publication metrics (i.e., number of publications and citation counts) to rewarding research practices that foster reproducibility, accountability, and transparency.

7.3.7   *Pre-registering studies*

To improve the clarity and meticulousness of a study's design, several journals, including the *Journal of Sex Research*, offer prospective authors the opportunity to pre-register their study's design (Sakaluk & Graham, 2022). With pre-registration a study's design can be scrutinised and suggested improvements can be made before data is collected and analysed (Center for Open Science, 2022). With improved research designs the prospect of successfully repeating an identical analysis of the dataset will increase (see 'Section 6.3.3 – Pre-registering a study's design' for more information about pre-registration of studies).

7.3.8   *Improving the readability of a study's methodology*

For a study to be reproducible, the steps and tools used to analyse the dataset need to be clearly explained. Failure to make the study's methodology easy to understand can result in an unsuccessful reproduction of the original study. Despite the need to clearly explain a study's methodology for reproducibility purposes, Plavén-Sigray et al. (2017) reported that over time the readability of scientific texts has decreased. They attributed this trend to increased usage of manuscript-specific jargon. As they explained:

> we showed that there is an increase in general scientific jargon over years. These general science jargon words should be interpreted as words which scientists frequently use in scientific texts, and not as subject specific jargon. This finding is indicative of a progressively increasing in-group scientific language ('science-ese').
>
> (Plavén-Sigray et al., 2017, p. 5)

To rectify this situation aspiring and current researchers should be given the opportunity to attend professional learning and development courses that focus on writing methodologies that are comprehensive, simple, and easy to understand. Such courses can complement current literature about how to write a manuscript for an academic journal (Barroga & Matanguihan, 2021; Forero et al., 2020).

7.3.9   *Improving the clarity of conference presentations*

To prevent the creation of irreproducible studies, researchers can improve their ability to present their academic work at conferences. Bautista et al. (2022) have provided ten suggestions that researchers can use to improve this skill. By improving this skill researchers should be able to reduce the prospect of their study being irreproducible because they would have clearly explained how they have analysed their data. Their ten suggestions were catalogued into one of three components. First, researchers should create opportunities to speak and should ask trusted colleagues for honest feedback about their presentation. Second, researchers should collaborate with others in their discipline and contribute to organising scientific meetings. Such encounters will give them the opportunity to present their research. Third, researchers should self-assess their presentation abilities. Such assessments can occur by learning about how other colleagues present their research and tailoring their own presentation to cater for the intended audience's interests and preliminary understanding of the field (see Figure 7.11) (Bautista et al., 2022).

7.3.10   *Requiring researchers to self-examine their previous research*

> In the current research environment, self-correction, or even just critical reconsideration of one's past work, is often disincentivized professionally. The opportunity costs of a self-correction are high; time spent on correcting past mistakes and missteps is time that cannot be spent on new research efforts, and the resulting self-correction is less likely to be judged a genuine scientific contribution. ... Researchers might also fear that a self-correction that exposes flaws in their work will damage their reputation and perhaps even undermine the credibility of their research record as a whole.
>
> (Rohrer et al., 2021, p. 1265)

As explained above, currently researchers are not incentivised to re-examine their previous academic work. Instead, they might resist engaging in such an act out of fear that their professional reputation will be damaged if they reveal that their previous academic work contained flaws. With a lack of self-examination, they are unable to learn from their past mistakes and

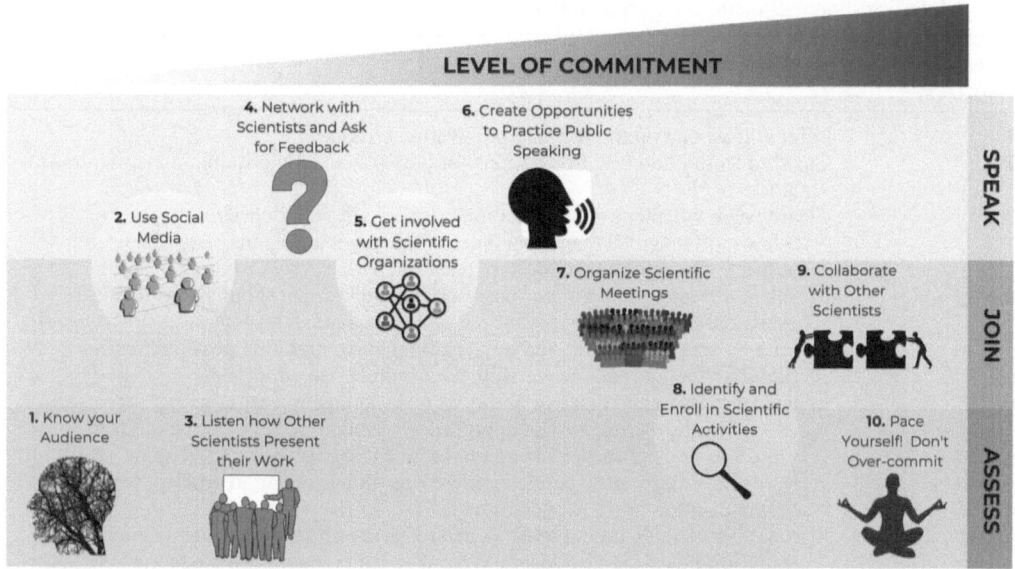

Figure 7.11 Ten simple rules for improving communication among scientists
Source: Bautista et al. (2022, p. 3).

improve their research skills and processes. Additionally, they are not able to easily master the skills needed to make their research reproducible since the incentive was to create highly cited research instead of research that could be reproduced (Fiala & Diamandis, 2017). Fiala and Diamandis (2017, p. 3) have described this situation as:

> Accountability in science is ad hoc. Researchers get credit for a publication well before enough time has passed for the scientific community to really know whether the paper has made a valuable contribution. No wonder that researchers bent on submitting a paper are obsessed with making the best possible case for its acceptance rather than illustrating its limitations. If researchers are forced to consider how well their paper will stand up five years hence, they will be more careful when doing the work and more critical in their analysis.

To compel authors to re-examine their own manuscripts after they have been published as a condition of publishing, authors could be required to publish a post-publication reflection of their published manuscript. To facilitate this process the editor could provide the author with a series of uncomplicated 'yes' or 'no' questions about the impact that their research has had on their chosen discipline. Additionally, they could also be asked a series of open-ended reflective questions, such as 'If you could repeat the study what design aspects would you change?' and 'What are some biases in your research (e.g., selection bias, sex and gender bias, and cultural bias)?' (Fiala & Diamandis, 2017). Finally, to motivate academics to publish opinion pieces about their past research Hardwicke et al. (2022) have proposed ten suggestions for editors and publishers (see Table 7.5).

7.4   Conclusion

To confirm that a study's results are correct the same results need to be created after repeating the same analytical processes on the same dataset. This chapter began with a definition of reproducibility and its distinction from the terms replicable, robust, and generalisable. The social and economic consequences of irreproducible research were then briefly explained, followed by an

Table 7.5 Policies journals could consider adopting to facilitate post-publication critiques

Recommendation number	Recommendation
1	Offer at least one option for post-publication critique.
2	Clearly identify and describe options for post-publication critique in instructions to authors.
3	Clearly state whether post-publication critique will be independently peer reviewed. Recognise that the authors of the target article may provide useful feedback, but cannot be considered neutral.
4	Facilitate expedient handling of post-publication critique submissions to ensure timely dissemination of relevant critique to research consumers.
5	Foster a culture of critique. Actively encourage and highlight post-publication critique to the journal's readership, for example, via editorials.
6	Enhance access to and discoverability of post-publication critique: (a) Tag post-publication critique with appropriate metadata so they can be indexed in third-party databases, websites and referencing software; (b) display prominent links to post-publication critique alongside target articles; (c) make post-publication critique open access.
7	Remove strict length, time-to-submit, and reference limits. Judge post-publication critique on a case-by-case basis and promote concise writing via editorial feedback.
8	Ensure transparent reporting of research articles (e.g., sharing of data, analysis code and materials, and adherence to reporting guidelines) to enable informed critique and debate.
9	Adopt a two-tier post-publication critique system. Tier one involves rapid publication of lightly moderated contributions on the journal's website (i.e., web comments). Tier two curates the most informative Tier one contributions and converts them to formal articles (letters) that become a permanent part of the scientific record and provide appropriate academic credit to their authors. For an example, two-tier system, see BMJ Rapid Responses.
10	Improve transparency and accountability by hiring an independent editor responsible for handling post-publication critique. Publish all editorial decisions related to post-publication critique, including the number submitted, rejected, and published.

Source: Hardwicke et al. (2022).

Note: In-text references have been removed.

overview of seven strategies that can improve the reproducibility of research, such as publishing datasets, implementing Open Science Badges, and incorporating reproducibility factors into the criteria for research funding. It is hoped that by integrating these and other strategies into existing research processes the ability to confirm published results will increase.

Additional readings

Contaxis, N., Clark, J., Dellureficio, A., Gonzales, S., Mannheimer, S., Oxley, P. R., Ratajeski, M. A., Surkis, A., Yarnell, A. M., Yee, M., & Holmes, K. (2022). Ten simple rules for improving research data discovery. *PLoS Computational Biology, 18*(2), e1009768. https://doi.org/10.1371/journal.pcbi.1009768

Drude, N., Martinez-Gamboa, L., Haven, T., Holman, C., Holst, M., Kniffert, S., McCann, S., Rackoll, T., Schulz, R., & Weschke, S. (2022). Finding the best fit for improving reproducibility: Reflections from the QUEST Center for Responsible Research. *BMC Research Notes, 15*(1), 270. https://doi.org/10.1186/s13104-022-06108-x

Hart, E. M., Barmby, P., LeBauer, D., Michonneau, F., Mount, S., Mulrooney, P., Poisot, T., Woo, K. H., Zimmerman, N. B., & Hollister, J. W. (2016). Ten simple rules for digital data storage. *PLoS Computational Biology, 12*(10), e1005097. https://doi.org/10.1371/journal.pcbi.1005097

Holst, M. R., Faust, A., & Strech, D. (2022). Do German university medical centres promote robust and transparent research? A cross-sectional study of institutional policies. *Health Research Policy and Systems, 20*(1), 39. https://doi.org/10.1186/s12961-022-00841-2

Macleod, M., & University of Edinburgh Research Strategy Group. (2022). Improving the reproducibility and integrity of research: What can different stakeholders contribute? *BMC Research Notes, 15*(1), 146. https://doi.org/10.1186/s13104-022-06030-2

Stewart, A. J., Farran, E. K., Grange, J. A., Macleod, M., Munafò, M., Newton, P., Shanks, D. R., & UKRN Institutional Leads. (2021). Improving research quality: The view from the UK Reproducibility Network institutional leads for research improvement. *BMC Research Notes, 14*(1), 458. https://doi.org/10.1186/s13104-021-05883-3

Stewart, S., Pennington, C. R., da Silva, G. R., Ballou, N., Butler, J., Dienes, Z., Jay, C., Rossit, S., Samara, A., & U. K. Reproducibility Network (UKRN) Local Network Leads. (2022). Reforms to improve reproducibility and quality must be coordinated across the research ecosystem: The view from the UKRN Local Network Leads. *BMC Research Notes, 15*(1), 58. https://doi.org/10.1186/s13104-022-05949-w

Turkyilmaz-van der Velden, Y., Dintzner, N., & Teperek, M. (2020). Reproducibility starts from you today. *Patterns (New York, N.Y.), 1*(6), 100099. https://doi.org/10.1016/j.patter.2020.100099

Wingen, T., Berkessel, J. B., & Englich, B. (2020). No replication, no trust? How low replicability influences trust in psychology. *Social Psychological and Personality Science, 11*(4), 454–463. https://doi.org/10.1177/1948550619877412

References

Adewumi, M. T., Vo, N., Tritz, D., Beaman, J., & Vassar, M. (2021). An evaluation of the practice of transparency and reproducibility in addiction medicine literature. *Addictive Behaviors, 112*, 106560. https://doi.org/10.1016/j.addbeh.2020.106560

American Psychiatric Association (APA) Journals. (2022). Retrieved from https://twitter.com/apa_journals/status/1030530992715563008?lang=hi

Barnett, A., Mewburn, I., & Schroter, S. (2019). Working 9 to 5, not the way to make an academic living: Observational analysis of manuscript and peer review submissions over time. *BMJ, 367*. https://doi.org/10.1136/bmj.l6460

Barroga, E., & Matanguihan, G. J. (2021). Creating logical flow when writing scientific articles. *Journal of Korean Medical Science, 36*(40), e275. https://doi.org/10.3346/jkms.2021.36.e275

Bautista, C., Alfuraiji, N., Drangowska-Way, A., Gangwani, K., de Flamingh, A., & Bourne, P. E. (2022). Ten simple rules for improving communication among scientists. *PLoS Computational Biology, 18*(6), e1010130. https://doi.org/10.1371/journal.pcbi.1010130

Borghi, J. A., & Van Gulick, A. E. (2021). Data management and sharing: Practices and perceptions of psychology researchers. *PLoS One, 16*(5), e0252047. https://doi.org/10.1371/journal.pone.0252047

Brito, J. J., Li, J., Moore, J. H., Greene, C. S., Nogoy, N. A., Garmire, L. X., & Mangul, S. (2020). Recommendations to enhance rigor and reproducibility in biomedical research. *GigaScience, 9*(6), giaa056. https://doi.org/10.1093/gigascience/giaa056

Center for Open Science. (2017). *Open Science Badges*. https://cos.io/our-services/open-science-badges/

Center for Open Science. (2022). *What Is Preregistration?* https://www.cos.io/initiatives/prereg

Devriendt, T., Borry, P., & Shabani, M. (2021). Factors that influence data sharing through data sharing platforms: A qualitative study on the views and experiences of cohort holders and platform developers. *PLoS One, 16*(7), e0254202. https://doi.org/10.1371/journal.pone.0254202

Diaba-Nuhoho, P., & Amponsah-Offeh, M. (2021). Reproducibility and research integrity: The role of scientists and institutions. *BMC Research Notes, 14*(1), 451. https://doi.org/10.1186/s13104-021-05875-3

Diong, J., Kroeger, C. M., Reynolds, K. J., Barnett, A., & Bero, L. A. (2021). Strengthening the incentives for responsible research practices in Australian health and medical research funding. *Research Integrity and Peer Review, 6*(1), 11. https://doi.org/10.1186/s41073-021-00113-7

Fiala, C., & Diamandis, E. P. (2017). Make researchers revisit past publications to improve reproducibility. *F1000Research, 6*. https://doi.org/10.12688/f1000research.12715.1

Fladie, I. A., Adewumi, T. M., Vo, N. H., Tritz, D. J., & Vassar, M. B. (2019). An evaluation of nephrology literature for transparency and reproducibility indicators: Cross-sectional review. *Kidney International Reports, 5*(2), 173–181. https://doi.org/10.1016/j.ekir.2019.11.001

Forero, D. A., Lopez-Leon, S., & Perry, G. (2020). A brief guide to the science and art of writing manuscripts in biomedicine. *Journal of Translational Medicine, 18*(1), 425. https://doi.org/10.1186/s12967-020-02596-2

Freedman, L. P., Cockburn, I. M., & Simcoe, T. S. (2015). The economics of reproducibility in preclinical research. *PLoS Biology, 13*(6), e1002165. https://doi.org/10.1371/journal.pbio.1002165

Grahe, J. E. (2014). Announcing open science badges and reaching for the sky. *The Journal of Social Psychology, 154*(1), 1–3. https://doi.org/10.1080/00224545.2014.853582

Hardwicke, T. E., Thibault, R. T., Kosie, J. E., Tzavella, L., Bendixen, T., Handcock, S. A., Köneke, V. E., & Ioannidis, J. (2022). Post-publication critique at top-ranked journals across scientific disciplines: A cross-sectional assessment of policies and practice. *Royal Society Open Science, 9*(8), 220139. https://doi.org/10.1098/rsos.220139

Hejblum, B. P., Kunzmann, K., Lavagnini, E., Hutchinson, A., Robertson, D. S., Jones, S. C., & Eckes-Shephard, A. H. (2020). *Realistic and Robust Reproducible Research for Biostatistics.* https://doi.org/10.20944/preprints202006.0002.v1

Ioannidis, J. P. (2012). Why science is not necessarily self-correcting. *Perspectives on Psychological Science: A Journal of the Association for Psychological Science, 7*(6), 645–654. https://doi.org/10.1177/1745691612464056

Johnson, A. L., Torgerson, T., Skinner, M., Hamilton, T., Tritz, D., & Vassar, M. (2020). An assessment of transparency and reproducibility-related research practices in otolaryngology. *The Laryngoscope, 130*(8), 1894–1901. https://doi.org/10.1002/lary.28322

Kanza, S., & Knight, N. J. (2022). Behind every great research project is great data management. *BMC Research Notes, 15*(1), 20. https://doi.org/10.1186/s13104-022-05908-5

Kidwell, M. C., Lazarević, L. B., Baranski, E., Hardwicke, T. E., Piechowski, S., Falkenberg, L. S., Kennett, C., Slowik, A., Sonnleitner, C., Hess-Holden, C., Errington, T. M., Fiedler, S., & Nosek, B. A. (2016). Badges to acknowledge open practices: A simple, low-cost, effective method for increasing transparency. *PLoS Biology, 14*(5), e1002456. https://doi.org/10.1371/journal.pbio.1002456

Kowalczyk, O. S., Lautarescu, A., Blok, E., Dall'Aglio, L., & Westwood, S. J. (2022). What senior academics can do to support reproducible and open research: A short, three-step guide. *BMC Research Notes, 15*(1), 116. https://doi.org/10.1186/s13104-022-05999-0

Kretser, A., Murphy, D., Bertuzzi, S., Abraham, T., Allison, D. B., Boor, K. J., Dwyer, J., Grantham, A., Harris, L. J., Hollander, R., Jacobs-Young, C., Rovito, S., Vafiadis, D., Woteki, C., Wyndham, J., & Yada, R. (2019). Scientific integrity principles and best practices: Recommendations from a scientific integrity consortium. *Science and Engineering Ethics, 25*(2), 327–355. https://doi.org/10.1007/s11948-019-00094-3

Leipzig, J., Nüst, D., Hoyt, C. T., Ram, K., & Greenberg, J. (2021). The role of metadata in reproducible computational research. *Patterns (New York, N.Y.), 2*(9), 100322. https://doi.org/10.1016/j.patter.2021.100322

Mede, N. G., Schäfer, M. S., Ziegler, R., & Weißkopf, M. (2021). The "replication crisis" in the public eye: Germans' awareness and perceptions of the (ir)reproducibility of scientific research. *Public Understanding of Science (Bristol, England), 30*(1), 91–102. https://doi.org/10.1177/0963662520954370

Minocher, R., Atmaca, S., Bavero, C., McElreath, R., & Beheim, B. (2021). Estimating the reproducibility of social learning research published between 1955 and 2018. *Royal Society Open Science, 8*(9), 210450. https://doi.org/10.1098/rsos.210450

Miyakawa, T. (2020). No raw data, no science: Another possible source of the reproducibility crisis. *Molecular Brain, 13*(1), 24. https://doi.org/10.1186/s13041-020-0552-2

Mozersky, J., McIntosh, T., Walsh, H. A., Parsons, M. V., Goodman, M., & DuBois, J. M. (2021). Barriers and facilitators to qualitative data sharing in the United States: A survey of qualitative researchers. *PLoS One, 16*(12), e0261719. https://doi.org/10.1371/journal.pone.0261719

Munafò, M. R., Chambers, C., Collins, A., Fortunato, L., & Macleod, M. (2022). The reproducibility debate is an opportunity, not a crisis. *BMC Research Notes, 15*(1), 43. https://doi.org/10.1186/s13104-022-05942-3

Okonya, O., Rorah, D., Tritz, D., Umberham, B., Wiley, M., & Vassar, M. (2020). Analysis of practices to promote reproducibility and transparency in anaesthesiology research. *British Journal of Anaesthesia, 125*(5), 835–842. https://doi.org/10.1016/j.bja.2020.03.035

Open Science Foundation. (2021, November 29). *View the Badges.* https://osf.io/tvyxz/wiki/1.%20 View%20the%20Badges/

Pexman, P. M. (2017). CJEP will offer open science badges. *Canadian Journal of Experimental Psychology = Revue canadienne de psychologie experimentale, 71*(1), 1. https://doi.org/10.1037/cep0000128

Plavén-Sigray, P., Matheson, G. J., Schiffler, B. C., & Thompson, W. H. (2017). The readability of scientific texts is decreasing over time. *eLife, 6*, e27725. https://doi.org/10.7554/eLife.27725

Rauh, S., Johnson, B. S., Bowers, A., Tritz, D., & Vassar, B. M. (2022). A review of reproducible and transparent research practices in urology publications from 2014 to 2018. *BMC Urology, 22*(1), 102. https://doi.org/10.1186/s12894-022-01059-8

Rauh, S., Torgerson, T., Johnson, A. L., Pollard, J., Tritz, D., & Vassar, M. (2020). Reproducible and transparent research practices in published neurology research. *Research Integrity and Peer Review, 5*, 5. https://doi.org/10.1186/s41073-020-0091-5

Rowhani-Farid, A., Aldcroft, A., & Barnett, A. G. (2020). Did awarding badges increase data sharing in *BMJ Open*? A randomized controlled trial. *Royal Society Open Science, 7*(3), 191818. https://doi.org/10.1098/rsos.191818

Rowhani-Farid, A., & Barnett, A. G. (2016). Has open data arrived at the British Medical Journal (BMJ)? An observational study. *BMJ Open, 6*(10), e011784. https://doi.org/10.1136/bmjopen-2016-011784

Sakaluk, J. K., & Graham, C. A. (2022). New Year, New Initiatives for *the Journal of Sex Research. Journal of Sex Research, 59*(7), 805–809. https://doi.org/10.1080/00224499.2022.2032571

Serghiou, S., Contopoulos-Ioannidis, D. G., Boyack, K. W., Riedel, N., Wallach, J. D., & Ioannidis, J. (2021). Assessment of transparency indicators across the biomedical literature: How open is open? *PLoS Biology, 19*(3), e3001107. https://doi.org/10.1371/journal.pbio.3001107

Sherry, C. E., Pollard, J. Z., Tritz, D., Carr, B. K., Pierce, A., & Vassar, M. (2020). Assessment of transparent and reproducible research practices in the psychiatry literature. *General Psychiatry, 33*(1), e100149. https://doi.org/10.1136/gpsych-2019-100149

Stewart, A. J., Farran, E. K., Grange, J. A., Macleod, M., Munafò, M., Newton, P., Shanks, D. R., & UKRN Institutional Leads. (2021). Improving research quality: The view from the UK Reproducibility Network institutional leads for research improvement. *BMC Research Notes, 14*(1), 458. https://doi.org/10.1186/s13104-021-05883-3

Tedersoo, L., Küngas, R., Oras, E., Köster, K., Eenmaa, H., Leijen, Ä., Pedaste, M., Raju, M., Astapova, A., Lukner, H., Kogermann, K., & Sepp, T. (2021). Data sharing practices and data availability upon request differ across scientific disciplines. *Scientific Data, 8*(1), 192. https://doi.org/10.1038/s41597-021-00981-0

van Elk, M., Rowatt, W., & Streib, H. (2018). Good dog, bad dog: Introducing open science badges. *The International Journal for the Psychology of Religion, 28*(1), 1–2. https://doi.org/10.1080/10508619.2018.1402589

Vazire, S., & Holcombe, A. O. (2021). Where are the self-correcting mechanisms in science? *Review of General Psychology, 26*(2), 212–223. https://doi.org/10.1177/10892680211033912

Wallach, J. D., Boyack, K. W., & Ioannidis, J. (2018). Reproducible research practices, transparency, and open access data in the biomedical literature, 2015–2017. *PLoS Biology, 16*(11), e2006930. https://doi.org/10.1371/journal.pbio.2006930

Walters, C., Harter, Z. J., Wayant, C., Vo, N., Warren, M., Chronister, J., Tritz, D., & Vassar, M. (2019). Do oncology researchers adhere to reproducible and transparent principles? A cross-sectional survey of published oncology literature. *BMJ Open, 9*(12), e033962. https://doi.org/10.1136/bmjopen-2019-033962

Wass, M. N., Ray, L., & Michaelis, M. (2019). Understanding of researcher behavior is required to improve data reliability. *GigaScience, 8*(5). https://doi.org/10.1093/gigascience/giz017

Wright, B. D., Vo, N., Nolan, J., Johnson, A. L., Braaten, T., Tritz, D., & Vassar, M. (2020). An analysis of key indicators of reproducibility in radiology. *Insights into Imaging, 11*(1), 65. https://doi.org/10.1186/s13244-020-00870-x

Appendix 7.1

Reproducibility practices on planet F345

An extremely intelligent humanlike race that resembles Homo sapiens lives on Planet F345 in the Andromeda galaxy. Here is the state of science on the planet in the year 3045268. Despite significant developments and diversity in scientific domains, most research is undertaken in a very small number of extremely popular subjects, each of which attracts the attention of tens of thousands of investigators and includes hundreds of thousands of articles. According to what we know from other civilisations in other galaxies, most of these fields are null fields – that is, fields where it has been demonstrated empirically that there are rare, if any, genuine discoveries. Thus, any discoveries are mostly due to random errors, bias, or both. In each of these null fields the results are simply estimates of the net bias occurring. Null domains include nutribogus epidemiology, pompompomics, social psycho junkology, and the various disciplines of brown cockroach research – brown cockroaches are regarded as providing a suitable model that can be easily applied to humanoids. Sadly, scientists on planet F345 are unaware that these are null domains and that they are squandering their time and lives by working in these disciplinary silos (Ioannidis, 2012).

New discoveries and the pursuit for statistically meaningful findings are taught to new scientists and researchers as the only things that really matter. In any respected university, hundreds of pre- and post-docs spend their entire careers using powerful computers to continuously mine data in enormous datasets. Anyone who discovers a spectacular Omega value (i.e., a figure determined from a statistical selection procedure) rushes to the chief researcher then reviews each article and only submits for peer review those that have the most striking findings. Universities are essentially managed by finance officers who are skilled at generating profits but have little to no knowledge or interest of science. Most university presidents, provosts, and deans are merely puppets employed solely for graduation speeches, other dull administrative events, and for expressing enthusiasm about new discoveries found at their institutions. Many research institutes' financial officers are hired after having had successful careers as real estate agents, store managers, or staff members of various corporate structures where they have demonstrated their ability to reduce costs and increase profits for their organisation. For researchers, their career progression is determined by their ability to provide increasingly radical, grandiose assertions and corresponding outcomes, which receive greater money even though nearly all of them are incorrect (Ioannidis, 2012).

On the planet F345, no studies are being replicated. Replication is viewed as an abhorrent activity fit only for fools who are only capable of copying what others have done, and it is certainly not recognised to be a legitimate branch of research. The most successful and frequent producers of false findings are members of the royal and national academies of science. Industry conducts many kinds of research and in some industries, like clinical medicine, this is nearly always the case. Again, the main objective is to produce exaggerated findings to patent new medical procedures, tests, and other technologies and enhance financial gain even though these findings are ineffective. Studies are created in such a way as to guarantee the production of findings with enough Omega values or, at the very least, to permit some modification to obtain Omega values that appear good (Ioannidis, 2012).

Even though there haven't been any significant discoveries in F345 for a long time, everyday announcements of new discoveries are broadcasted to average residents by the media. Most nations on F345 are typically against critical thinking and enquiry. Free and critical thinking has become perceived as a nuisance courtesy of free markets that obliterated democratic constitutions and freedom of thought. As a result, totalitarian countries with a lack of freedom of expression or massive social inequalities – one of the most striking being gender inequalities against men – have the largest salaries for scientists and the most sophisticated research infrastructure. Science thrives in environments where freedom of thought and critical questioning is constrained since on F345 such acts (including, of course, efforts to replicate claimed findings) are deemed antithetical to good science (Ioannidis, 2012).

Source: Ioannidis (2012).

Appendix 7.2

Open Science Badges Published by the Center for Open Science

Open Science Badge icon	Description of the Open Science Badge icon
	The **Open Data** badge is awarded when digitally-shareable data necessary to reproduce the reported results are publicly available. ***Criteria*** 1 Digitally-shareable data are publicly available on an open-access repository. The data must have a persistent identifier and be provided in a format that is time-stamped, immutable, and permanent (e.g., university repository, a registration on the Open Science Framework, or an independent repository at www.re3data.org). 2 A data dictionary (for example, a codebook or metadata describing the data) is included with sufficient description for an independent researcher to reproduce the reported analyses and results. Data from the same project that are not needed to reproduce the reported results can be kept private without losing eligibility for the Open Data Badge. 3 An open licence allowing others to copy, distribute, and make use of the data while allowing the licensor to retain credit and copyright as applicable. Creative Commons has defined several licenses for this purpose, which are described at www.creativecommons.org/licenses. CC0 or CC-BY is strongly recommended. ***State of Data Notations*** Specification of open data is complicated by the fact that raw, collected data may be processed prior to conducting the reported analyses leading to derived, constructed data based on the raw data. For example, the raw data might be ten survey responses and the constructed data might be the mean score of those ten responses. Open data badges assume that at least raw data are available. If only derived, constructed data are available (i.e., the data used to conduct the reported analyses), it is denoted with the badge. Sharing derived datasets must include descriptions of how the data were constructed or, even better, provide the code used to construct the data.

(Continued)

(Continued)

Open Science Badge icon	Description of the Open Science Badge icon

Specification of open data is also complicated by the fact that the data underlying the reported analyses may be a subset of the data collected for the study. Open data badges assume that all collected data are made available. If only the subset of data used to conduct the reported analyses is available, it is denoted with the badge. Sharing reported subsets must include descriptions of how the data were reduced from the complete dataset or, even better, provide the code used to reduce the dataset.

Specification of open data explicitly excludes data that compromises confidentiality or anonymity of human participants. If access to identifying data is necessary to reproduce the reported analyses, then the report is not eligible for an open data badge

The **Open Materials** badge is earned by making publicly available the components of the research methodology needed to reproduce the reported procedure and analysis.

Criteria

1 Digitally-shareable materials are publicly available on an open-access repository. The materials must have a persistent identifier and be provided in a format that is time-stamped, immutable, and permanent (e.g., university repository, a registration on the Open Science Framework, or an independent repository at www.re3data.org).

2 Infrastructure, equipment, biological materials, or other components that cannot be shared digitally are described in sufficient detail for an independent researcher to understand how to reproduce the procedure.

3 Sufficient explanation for an independent researcher to understand how the materials relate to the   reported methodology.

Pre-registered

The Pre-registered badge is earned for having a pre-registered design. A pre-registered design includes: (1) Description of the research design and study materials including planned sample size, (2) Description of motivating research question or hypothesis, (3) Description of the outcome variable(s), and (4) Description of the predictor variables including controls, covariates, independent variables (conditions). When possible, the study materials themselves are included in the pre-registration.

Criteria for earning the pre-registered badge on a report of research are:

1 A public date-time stamped registration is in an institutional registration system (e.g., ClinicalTrials.gov, Open Science Framework, AEA Registry, and EGAP).

(Continued)

Open Science Badge icon	Description of the Open Science Badge icon

2 Registration pre-dates the intervention.

3 Registered design corresponds directly to reported design.

4 Full disclosure of results in accordance with registered plan.

Badge eligibility does not restrict authors from reporting results of additional analyses. Results from pre-registered analyses must be distinguished explicitly from additional results in the report. Notations may be added to badges. Notations qualify badge meaning: TC, or Transparent Changes, means that the design was altered but the changes and rationale for changes are provided. DE, or Data Exist, means that (2) is replaced with "registration postdates realisation of the outcomes, but the authors have yet to inspect or analyse the outcomes."

Pre-registered+Analysis Plan

The Pre-registered+Analysis Plan badge is earned for having a pre-registered research design (described above) and an analysis plan for the research and reporting results according to that plan. An analysis plan includes specification of the variables and the analyses that will be conducted. Guidance on construction of an analysis plan is below.

Criteria for earning the pre-registered+analysis plan badge on a report of research are:

1 A public date-time stamped registration is in an institutional registration system (e.g., ClinicalTrials.gov, Open Science Framework, AEA registry, and EGAP).

2 Registration pre-dates the intervention.

3 Registered design and analysis plan corresponds directly to reported design and analysis.

4 Full disclosure of results in accordance with the registered plan.

Notations may be added to badges. Notations qualify badge meaning: TC, or Transparent Changes, means that the design or analysis plan was altered but the changes are described and a rationale for the changes is provided. Where possible, analyses following the original specification should also be provided. DE, or Data Exist, means that (2) is replaced with "registration postdates realisation of the outcomes, but the authors have yet to inspect or analyse the outcomes".

Guidance on Analysis Plans

Procedures

- What is your planned sample size?
- If applicable, how many individual units and how many clusters?
- If you are conducting a randomised control trial or experimental study, how will you randomise?

(Continued)

Open Science Badge icon	Description of the Open Science Badge icon
	• At what level will you randomise (individual or cluster level)?

Exclusions

• What conditions will lead to data being excluded?
Variable Construction

• If your predictor variable(s) are not from a single question or measure, how will they be constructed?
• If your outcome variable(s) are not from a single question or measure, how will they be constructed?

Tests or models

• What is the quantity you intend to estimate?
• What is the unit of analysis (if applicable)?
• What statistical model(s) will you use to test your hypothesis? Please include the type of model (e.g., ANOVA, regression, and SEM) as well as the specification of the model (e.g., what variables will be included and how they will be included).
• If you are comparing multiple conditions or testing multiple outcomes and/or hypotheses, how will you account for this?

In addition, the researcher will be invited to pre-specify procedures that will be used in the event of foreseeable problems (e.g., attrition, noncompliance, and failure to enrol the planned number of subjects) that routinely afflict certain kinds of studies. |

Source: Open Science Foundation (2021).

8 Human Research Ethics Committees and Metascience

Abstract

This chapter begins with an overview about the historical events that resulted in the creation of Human Research Ethics Committees (HRECs), which are also known as institutional review boards, independent ethics committees, ethical review boards, or research ethics boards. Seven issues with how HRECs operate are then outlined. For example, the lack of a retrospective ethics review session within the process of obtaining ethics approval. Seven solutions to these operational issues are also described. The intention of these strategies is to improve the likelihood of applicants obtaining ethics approval and to make the tasks that members of HRECs perform easier.

Keywords: Ethics Education; Human Research Ethics Committees (HRECs); Prospective Ethics Reviews; Retrospective Ethics Reviews

Key points

- Human Research Ethics Committees (HRECs) approve a study and then the applicant conducts the study and is not obligated to explain to the HREC any ethical issues that arose during the study. This lack of reporting means that a HREC is unable to anticipate possible ethical issues that might arise with other applications. A retrospective ethics review could rectify this design flaw.
- Some members of HRECs do not understand the ethics application that they are examining. To assist such members, they can receive training and education about the concepts in the ethics proposal and/or applicants for ethics approval can make interactive ethics applications in which they explain their project and answer any questions from members of the HREC.

8.1 The creation of Human Research Ethics Committees

There are two historical events that have possibly contributed to the development of HRECs. First, the inhumane experimentation on human subjects by the Nazis during the Second World War motivated the international community to develop and enforce policies and regulations that protected the rights of human subjects (Stevenson et al., 2015). Second, prior to the development of HRECs, a series of ethically questionable psychology studies occurred, such as Zimbardo's Stanford Prison Experiment, the Tuskegee Syphilis Study, the Willowbrook hepatitis experiments, and Milgram's obedience experiments (Spellecy & Busse, 2021; Stevenson et al., 2015). These types of studies shifted the somewhat nonchalant attitudes that society had

DOI: 10.4324/9781003510376-8

towards researchers conducting experiments on humans up to that point in time. Although the creation of HRECs is a welcomed development, especially in terms of protecting human participants, the question 'Are HRECs operating efficiently and effectively protecting the safety of participants in research?' continues to be the subject of much debate and commentary.

Concerns about the performance of HRECs in safeguarding the interests of participants is not a novel phenomenon. More than two decades ago Savulescu and colleagues (1996) published the article 'Are research ethics committees behaving unethically? Some suggestions for improving performance and accountability'. With the intention of improving the HREC's ability to protect the interests of participants within this article they proposed five strategies, which were:

1 Before the HREC approves an ethics proposal the applicant should give the HREC a systematic review of the existing literature,
2 Applicants should give potential participants a summary of the relevant literature review and explain how their research will contribute to knowledge,
3 Clinical trials should be registered as a condition of approval,
4 As a condition of approval, the HREC should require a commitment from the researchers that the results from their study will be made publicly accessible, and
5 Previously approved ethics proposals should be audited to determine the effectiveness of the HREC in identifying any potential ethical problems (Savulescu et al., 1996).

Since the recommendations suggested by Savulescu and colleagues, other scholars have critiqued the effectiveness of HRECs in safeguarding the welfare and interests of participants. Issues impacting the effectiveness and efficacy of HRECs will now be presented and then the remainder of this chapter will examine several strategies that can improve the efficiency and purpose of HRECs. The aim of these strategies is greater participant safety, less public expenditure being spent on compensating participants who were subjected to ethical violations, and improved efficiency of HRECs.

8.2   Operational issues with Human Research Ethics Committees

8.2.1   *Providing ethics training to applicants*

Sometimes HRECs are burdened with re-examining amended ethics applications because the initial applications contained errors. The amendments requested by the HREC can range from correcting grammatical and spelling errors to re-writing and clarifying complete sections of the ethics proposal (Brandenberg et al., 2021). HRECs can provide applicants with training about how to successfully complete a HREC application to reduce this burden. This sentiment was expressed by Davis and colleagues (2021) who proposed that HRECs should furnish applicants with clear and comprehensive guides to ethics application processes and design curricula that improves a student's understanding of such processes. Similarly, Brandenberg and colleagues (2021, p. 356) claimed:

> Possible ways of improving this range from very simple, for example, checklists of administrative requirements, to more complex, like improved ethics training for researchers, or more specific guidance from HRECs or national ethics bodies (for instance, standard guidance on data management/storage for the institution).

8.2.2 *Educating members of Human Research Ethics Committees to examine ethics applications*

> When I served on an ethics committee one of the applications that came before us was for a repetition of a study that had been done elsewhere using a particular survey instrument. And the members of the committee wanted to change the wording of some of the items. I argued strongly that, you know, you couldn't do that. They were using standardized instruments, you can't do that. But he insisted that it should be changed to, you know, to some other... not all members of the ethics committee at that time were fully conversant with their role, I don't think. We're not going on an ethics committee to redesign the study. Maybe to advise on methodology... You don't allow bad research through. Badly designed I mean. But to tinker with the methodology at the edges I think is outside their remit.
>
> (Guillemin et al., 2012, p. 42)

As the quote above illustrated, some members of HRECs are ill-equipped to proficiently examine the contents of an ethics proposal (Ainembabazi et al., 2021; Barnard et al., 2021; Pysar et al., 2021). Barnard and colleagues (2021) investigated the views and experiences of members of HRECS regarding evaluating ethics proposals about suicide. They concluded that some HREC members might be inexperienced in dealing with suicide-related study applications. Such inexperience may result in them approving ethics proposals that contain insufficient safeguards for participants who have attempted suicide, or not approving studies that are well thought out with good safeguards, due to a fear of negatively impacting vulnerable participants. Pysar and colleagues (2021) examined the competency of HREC members to evaluate ethics applications for genomic research. They concluded that even HREC members who had completed genomics education modules expressed uncertainty about examining the merits of ethics applications about genomic research. This leads one to conclude that where HREC members have not had genomic research specific experience and/or education could make decisions that in retrospect might be inappropriate.

To resolve such issues the applicant could nominate a subject matter expert familiar with the research that members of the HREC could contact to provide them with the necessary contextual information. It is hypothetically possible that the subject matter expert might either sabotage or inappropriately endorse the applicant's submission. To avoid any bias the subject matter expert should be unaware of the applicant's identity and must not be allowed to vote on the applicant's submission.

8.2.3 *Interactive ethics presentations*

Typically, to obtain ethics approval an applicant must complete a series of standardised forms that explain their proposed research and any perceived potential ethical issues. The HREC then examines this paperwork and decides if the study complies with ethical standards and should proceed or not. However, this approach contains two flaws. First, any additional information that the HREC requires to decide if the study is ethically appropriate must be requested from the applicant after the HREC meeting, which can cause delays with approval. Second, sometimes members of the HREC misunderstand the applicant's ethics proposal. For example, a respondent in Guillemin et al.'s (2012, p. 44) study claimed:

> With the ethics committee, I had the chance to go and talk to the person who was reviewing the application and explain things. Part of why it happened was the student wrote the ethics application with me...the way things were expressed was perhaps not as clear as they could be. Also the person reviewing the ethics application probably had, I don't want to say limited,

but was viewing the research methodology in a particular way. So when I was able to meet with this person and explain the methodology we were right.

To offset the occurrence of such misunderstandings and to provide members of the HREC with additional information, which should quicken the application process, applicants should be invited to the HREC meeting. At such meetings applicants could give an oral presentation about their study and answer any questions. This presentation would also assist members of the HREC who prefer to receive information orally. To guide the applicants the HREC should outline the topics that they can explain, such as recruitment strategies, data archival protocols, and how they will preserve the participant's confidentiality. Participants can also draw inspiration from presentation guidelines to deliver an effective presentation.

8.2.4   *Retrospective ethics reviews*

A 'prospective ethics review' is the typical pathway for a HREC to grant ethics approval. With this process a study is designed, an ethics application is drafted, reviewed, submitted to the HREC, and then once the ethics approval is given the study commences. Although this approach is traditional practice, Dawson and colleagues (2019) proposed that it has two significant limitations, which are:

> First, as currently practiced, research ethics has become for some a 'tick box' exercise to get over the 'hurdle' of ethics approval. This fails to capture much of what is important in ethics and does not promote careful reflection on the ethical issues involved. Second, the current approach tends to be rules-based and we argue that research ethics should go beyond this to develop people's capacity to be sensitive to the relevant moral features of their research, their ethical decision-making skills and their integrity.
>
> (Dawson et al., 2019, p. 1)

Dawson and colleagues have suggested that to mitigate these two limitations the HREC should make a condition of granting ethics approval at the prospective review stage that once a study is finished the HREC and the researcher should conduct a retrospective ethics review. Dawson and colleagues have claimed that retrospective ethics reviews can give researchers the opportunity to reflect upon their research and:

> would allow discussion about the harms and benefits that 'actually' arose and how those compared with what was predicted. We acknowledge that such reflection does occur during and after some research, but is largely informal and not disseminated in ways that allow others to learn from such experiences.
>
> (Dawson et al., 2019, p. 3)

An internally or externally published retrospective ethics review can also uncover ethical issues that the prospective ethics review failed to anticipate and when this occurs future prospective ethics reviews may be improved. A five-stage cyclical process can be used to conceptualise the benefits that retrospective ethics reviews give prospective ethics reviews. Stage one involves the researcher preparing an ethics application that will be examined during the prospective ethics review. In stage two this ethics application is submitted to and reviewed by the HREC and is either passed unamended or approved after modifications have been implemented. For stage three the study is conducted in accordance with the ethics application that was approved in stage two. At stage four with the intention of noting any ethical issues that arose the researcher reviews how the

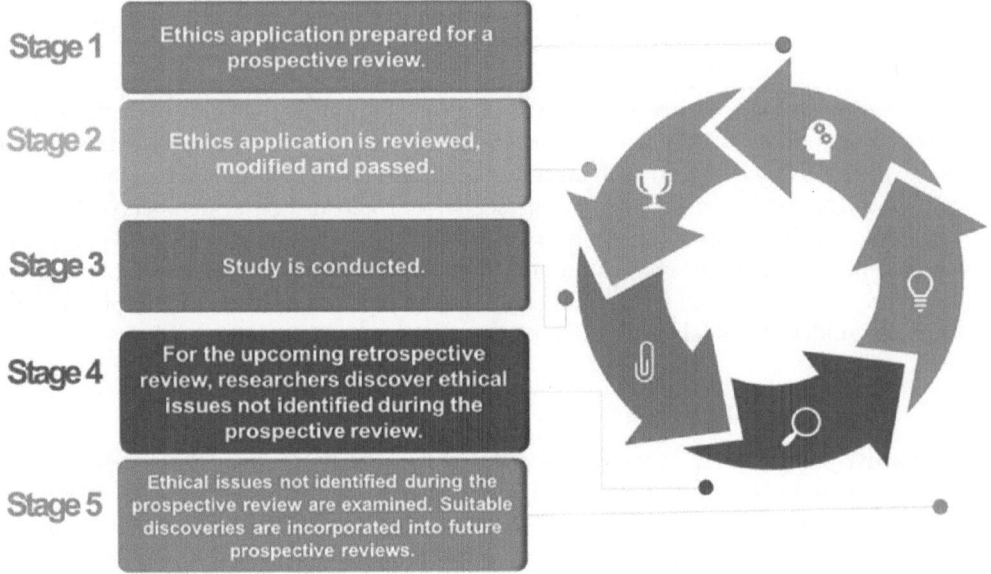

Figure 8.1 Five-stage feedback loop about the benefits that retrospective ethics reviews can give prospective ethics reviews

study was conducted and if they identified any anomalies during the prospective ethics review stage. For stage five the researcher and members of the HREC conduct the retrospective ethics review. During this stage there is a particular focus on any ethical issues not identified during the prospective review stage. These issues are examined and noted so that suitable discoveries can be incorporated into future prospective ethics reviews. For example, during a retrospective ethics review a researcher may uncover deficiencies that occurred during the consent process. With this discovery, in the future prospective ethics reviews may change the way that they evaluate the appropriateness of consent forms (Coleman & Bouësseau, 2008) (see Figure 8.1).

8.2.5   *Participant feedback to ethics committees*

To obtain ethics approval from a HREC a prospective applicant typically identifies from a predefined list any potential ethical concerns and/or explain any potential ethical concerns that were not predefined. Applicants also must explain how they will minimise any ethical concerns. In contrast, aside from receiving complaints about the conduct of researchers, HRECs never approach former participants and ask them if they believed that the study was conducted ethically. HRECs are consequently unaware of ethical violations against participants who have not been able, for whatever reason, to lodge a complaint to the HREC. The ramifications of HRECs not approaching former participants are that participants who are unable to advocate for themselves remain the victims of research conducted unethically and researchers who performed ethical violations remain undetected.

8.2.6   *Creating consistent policies for Health Research Ethics Committees*

Within some nations, such as Australia, there are inconsistent policies relating to HRECs (Dudi-Venkata et al., 2021). Dudi-Venkata and colleagues (2021) reported that one-third of their respondents received inconsistent advice and documentation between HRECs in Australia.

They suggested that across Australia ethics processes varied and that this variability contributed to inconsistent decisions and outcomes. To avoid this result, they recommended a centralised application process that would enhance collaborative research efforts. To achieve this objective national research organisations, such as the *National Institutes of Health* in the United States and the *Australian Research Council,* should provide ethics applications proforma that all HRECs must adopt unaltered. Such standardisation could provide applicants with a single form to be submitted to any number of HRECs when they need multiple approvals.

8.2.7 *Reducing HREC application rejection rates*

Typically, HRECs reject ethics applications that are incomplete or flawed. When members of HRECs encounter substandard ethics applications they use time and resources to draft recommendations for the application's improvement. Using content analysis, Brandenburg and colleagues (2021) audited the correspondence that a single HREC in Australia sent applicants regarding modifications to their ethics proposal. They examined 24 submissions, containing 355 distinct request components. The most frequent modification that applicants received was requests to correct administrative mistakes, which occurred in every submission. Altering the Participant Information and Consent Form was the second most requested modification, which occurred in 79% of submissions examined. In more than 50% of examined submissions, the HREC requested that the applicant clarify details about data collection and study procedures, general ethical considerations, recruitment and consent, setting or patient pool, research design and methodology, and data management and security. Regarding the overall feedback by the HREC, 44% of them were direct corrections or specific requests for changes, 42% were requests for additional information or clarification of already-provided information, and 14% were the HREC expressing concerns about a study component without directly proposing a change.

To improve the overall quality of ethics applications sent to HRECs, which will reduce the resources needed to explain to applicants why their ethics application was not approved, it is essential that applicants receive appropriate training and guidance about completing ethics applications. Additionally, applicants can use checklists to ensure that their ethics proposal is complete and of high-quality prior to its submission (Brandenburg et al., 2021) (see Appendix 8.1).

8.3 Conclusion

The focus of this chapter was explaining the shortfalls in some HREC processes and how these limitations can be mitigated. It began with the historical factors that resulted in the creation of HRECs. Among other suggestions, it then explained the advantages associated with educating members of HRECs about the technical concepts within ethics applications and the merits of implementing a retrospective ethics review. With the intention of reducing the amount of inferior ethics applications that HRECs reject or request modifications this chapter concluded with the suggestion that applicants can use an ethics checklist before they submit their ethics application to the HREC for examination.

Additional readings

Brindley, R., Nolte, L., & Nel, P. W. (2020). We were in one place, and the ethics committee in another: Experiences of going through the research ethics application process. *Clinical Ethics, 15*(2), 94–103. https://doi.org/10.1177/1477750920903454

Hokke, S., Hackworth, N. J., Bennetts, S. K., Nicholson, J. M., Keyzer, P., Lucke, J., Zion, L., & Crawford, S. B. (2020). Ethical considerations in using social media to engage research participants: Perspectives of Australian researchers and ethics committee members. *Journal of Empirical Research on Human Research Ethics: JERHRE, 15*(1–2), 12–27. https://doi.org/10.1177/1556264619854629

Lynch, H. F., Eriksen, W., & Clapp, J. T. (2022). "We measure what we can measure": Struggles in defining and evaluating institutional review board quality. *Social Science & Medicine (1982), 292,* 114614. https://doi.org/10.1016/j.socscimed.2021.114614

Petillion, W., Melrose, S., Moore, S. L., & Nuttgens, S. (2017). Graduate students' experiences with research ethics in conducting health research. *Research Ethics, 13*(3–4), 139–154. https://doi.org/10.1177/1747016116677635

Scherer, A., Alt-Epping, B., Nauck, F., & Marx, G. (2019). Team members perspectives on conflicts in clinical ethics committees. *Nursing Ethics, 26*(7–8), 2098–2112. https://doi.org/10.1177/0969733019829857

References

Ainembabazi, P., Castelnuovo, B., Okoboi, S., Arinaitwe, W. J., Parkes-Ratanshi, R., & Byakika-Kibwika, P. (2021). A situation analysis of competences of research ethics committee members regarding review of research protocols with complex and emerging study designs in Uganda. *BMC Medical Ethics, 22*(1), 132. https://doi.org/10.1186/s12910-021-00692-6

Barnard, E., Dempster, G., Krysinska, K., Reifels, L., Robinson, J., Pirkis, J., & Andriessen, K. (2021). Ethical concerns in suicide research: Thematic analysis of the views of human research ethics committees in Australia. *BMC Medical Ethics, 22*(1), 41. https://doi.org/10.1186/s12910-021-00609-3

Brandenburg, C., Thorning, S., & Ruthenberg, C. (2021). What are the most common reasons for return of ethics submissions? An audit of an Australian health service ethics committee. *Research Ethics, 17*(3), 346–358. https://doi.org/10.1177/1747016121999935

Coleman, C. H., & Bouësseau, M. C. (2008). How do we know that research ethics committees are really working? The neglected role of outcomes assessment in research ethics review. *BMC Medical Ethics, 9,* 6. https://doi.org/10.1186/1472-6939-9-6

Davis, K., Tan, L., Miller, J., & Israel, M. (2022). Seeking Approval: International Higher Education Students' Experiences of Applying for Human Research Ethics Clearance in Australia. *Journal of Academic Ethics, 20*(3), 421–436. https://doi.org/10.1007/s10805-021-09425-1

Dawson, A., Lignou, S., Siriwardhana, C., & O'Mathúna, D. P. (2019). Why research ethics should add retrospective review. *BMC Medical Ethics, 20*(1), 68. https://doi.org/10.1186/s12910-019-0399-1

Dudi-Venkata, N. N., Cox, D., Marson, N., Tan, L., Pockney, P., Muralidharan, V., Watson, D. I., Richards, T., & Clinical Trials Network Australia New Zealand (CTANZ). (2021). Variation in Human Research Ethics Committee and governance processes throughout Australia: A need for a uniform approach. *ANZ Journal of Surgery, 91*(11), 2263–2268. https://doi.org/10.1111/ans.16842

Guillemin, M., Gillam, L., Rosenthal, D., & Bolitho, A. (2012). Human research ethics committees: Examining their roles and practices. *Journal of Empirical Research on Human Research Ethics, 7*(3), 38–49. https://doi.org/10.1525/jer.2012.7.3.38

Pysar, R., Wallingford, C. K., Boyle, J., Campbell, S. B., Eckstein, L., McWhirter, R., Terrill, B., Jacobs, C., & McInerney-Leo, A. M. (2021). Australian human research ethics committee members' confidence in reviewing genomic research applications. *European Journal of Human Genetics, 29*(12), 1811–1818. https://doi.org/10.1038/s41431-021-00951-5

Savulescu, J., Chalmers, I., & Blunt, J. (1996). Are research ethics committees behaving unethically? Some suggestions for improving performance and accountability. *BMJ (Clinical research ed.), 313*(7069), 1390–1393. https://doi.org/10.1136/bmj.313.7069.1390

Spellecy, R., & Busse, K. (2021). The history of human subjects research and rationale for institutional review board oversight. *Nutrition in Clinical Practice: Official Publication of the American Society for Parenteral and Enteral Nutrition, 36*(3), 560–567. https://doi.org/10.1002/ncp.10623

Stevenson, F. A., Gibson, W., Pelletier, C., Chrysikou, V., & Park, S. (2015). Reconsidering 'ethics' and 'quality' in healthcare research: The case for an iterative ethical paradigm. *BMC Medical Ethics, 16,* 21. https://doi.org/10.1186/s12910-015-0004-1

Appendix 8.1

Human Research Ethics Committee checklist

Item	Checked	Not applicable
Have all documents been attached and submitted to the HREC application?	□	□
Have all typographical and grammatical errors in the HREC application and other attached documents been identified and corrected?	□	□
Has the relevant document name, version, page numbers, and date been placed in the footer on each page of each relevant document?	□	□
Are details between each document consistent? For example, the participant recruitment flyer and HREC ethics application should both state that the participant will be interviewed once for between 30 and 60 minutes.	□	□
Are the contact details for all members of the research team included in the HREC application?	□	□
Are the contact details for all members of the research team corrected in the HREC application?	□	□
Are all attached documents cited in the HREC application? For example, a citation in the HREC application should be made to the attached interview questions.	□	□
Before submission have all corrections of the HREC been made?	□	□
Has all unnecessary technical jargon and complex wording been removed?	□	□
Does the HREC application contain sufficient details about:	□	□
• the study's purpose?	□	□
• who in the research team will have access to de-identified information about the participant?	□	□
• who can the participant contact if the study causes them distress?	□	□
• the study's procedures?	□	□
• how much funding, resources, and/or in-kind support has been allocated to the study and who has provided this support?	□	□
• the support that will be given to participants before, during, and after the study?	□	□
• how health information and incidental findings will be explained to participants? For example, if the study involved the participant's brain being scanned and the brain scan revealed a tumour will the research team tell the participant and provide counselling and support.	□	□
• the inclusion and exclusion criteria for participants?	□	□
• if and why a waiver of consent was needed?	□	□
• the number of participants expected to be in the sample?	□	□
• where the study will be conducted? For example, at a university, research institute, or a clinic.	□	□

(Continued)

(Continued)

Item	Checked	Not applicable
• how biases will be mitigated? For example, selection and confirmation bias.	☐	☐
• the role of each member of the research team?	☐	☐
• data and/or sample storage.	☐	☐
• how the collected data will be analysed?	☐	☐
• how the results will be validated?	☐	☐
• the potential risks and burdens to participants?	☐	☐
• what the participants are expected to do in the study?	☐	☐
• how will the participant's identity remain confidential?	☐	☐
• how potential participants will be recruited?	☐	☐

9 Final remarks

As we explained in the first chapter, there have been several high-profile academic hoaxes that have exposed the inherent flaws in our current processes used to create research. Such stunts tend to divide those in the academic community. On the one hand, some have seized upon these hoaxes to justify a 'call for change' to the ways that we produce research. While, on the other hand, others have condemned these hoaxers, perhaps in the hope that their disapproval will dissuade others from performing similar stunts. For transparency, we applaud and congratulate Helen Pluckrose, James Lindsay, Peter Boghossian, and Alan Sokal for the hoaxes that they performed. Their actions serve to remind us that the production of research is flawed and that everyone in the research community, such as editors and peer reviewers, should continuously examine their actions and seek better ways of creating and disseminating research.

In our opinion, while the attempts of hoaxers are noble, since they raise awareness that the tools that we use to create research are flawed, others have unethical motives. Since the development of knowledge, there have been dishonest researchers who have published false research with the intention of enhancing their careers and swindling millions of research dollars, allocated in good faith by taxpayers, for personal financial gain. One pertinent example of recent academic misconduct is the allegation that Sylvain Lesné falsified images of the brains of Alzheimer's participants in his 2006 article published by *Nature* called 'A specific amyloid-β protein assembly in the brain impairs memory'. As of 2022, this article has been accessed more than 47,000 times, has been cited 2,280 times, and has an Altmetric score of 1,215 (Lesné et al., 2006). Although these allegations remain unproven, as of 14 July 2022 *Nature* took the cautious step of placing a disclaimer on the article warning readers that:

> **14 July 2022** Editor's Note: The editors of Nature have been alerted to concerns regarding some of the figures in this paper. Nature is investigating these concerns, and a further editorial response will follow as soon as possible. In the meantime, readers are advised to use caution when using results reported therein.

Our widely accepted understanding of Alzheimer's disease could be profoundly undermined if the allegations against Lesné et al. are true. Their findings support the popular but contentious 'amyloid theory' of Alzheimer's disease, which proposes that Aβ clumps in brain tissue, known as 'plaques', are the primary cause of this disease. The integrity of the amyloid theory is undermined if they have falsified the images in their article. If the allegations against Lesné et al. are true, the taxpayer finances used to produce this research would have been wasted and will, most likely, never be recovered. For those with Alzheimer's disease and their families, they have unnecessarily suffered over the last decade while researchers, acting with the best of intentions, have unknowingly pursued a cure for Alzheimer's disease based on specious research.

DOI: 10.4324/9781003510376-9

Looking at the current state of science and our society we were motivated to write this book for two reasons. First, considering the damage that academic charlatans can, and have done, to taxpayer investment in research we hope that the concepts in this book can be guardrails that safeguard the discovery of new knowledge. Second, we hope that this book will help protect and enhance the reputation of research in an age when a small vocal minority of the public espouse sceptical, suspicious, and distrusting views of researchers and verified knowledge. In our opinion, metascience is an invaluable discipline to achieving our desire for a society in which the creation of knowledge is an honest profession that the public can trust. Due to its purpose and nature, metascience forces researchers to stop and make a conscious effort to evaluate their actions, the factors that are in their workplaces that undermine new breakthroughs, and the effectiveness of the tools that they use to discover new knowledge.

At a fundamental level, metascience challenges our ingrained assumptions about how we should create research. For example, countless generations of academics have customarily destroyed their datasets once they have published their study. Within academia, the act of destroying datasets is habitual because each generation of academic supervisors teaches the next this protocol. As each generation learns this practice, it becomes normalised, instinctual, and rarely questioned. Only when problems arise with this practice, as is the case that the results in some disciplines cannot be confirmed, that researchers turn to metascience for guidance. *Reproducibility*, a prominent concept in metascience, challenges this established practice of destroying datasets because it contends that datasets need to be preserved so that the study's results can be confirmed.

In closing, like other disciplines, metascience will continue to evolve and change as new ways of creating and disseminating knowledge are discovered. However, regardless of its evolutionary journey, we hope that this book has given you a fundamental understanding and appreciation of metascience, an essential discipline that safeguards the creation of reputable research.

Reference

Lesné, S., Koh, M. T., Kotilinek, L., Kayed, R., Glabe, C. G., Yang, A., Gallagher, M., & Ashe, K. H. (2006). A specific amyloid-beta protein assembly in the brain impairs memory. *Nature, 440*(7082), 352–357. https://doi.org/10.1038/nature04533

Index

Note: **Bold** page numbers refer to tables and *italic* page numbers refer to figures.